French Aces of WWI

Volume 1: A–G

Alexandr Kazakov & Anton Zhukov

Acknowledgements
For our aviation books in print, please see our website at: **www.aeronautbooks.com**.
You may contact me at **jherris@me.com**

ISBN: **978-1-953201-12-6**

Design and layout: Jack Herris
Cover design: Aaron Weaver
Digital photo editing: Jack Herris

Aeronaut Books
Books for Enthusiasts by Enthusiasts
www.aeronautbooks.com

Table of Contents

Foreword

There are a huge number of publications devoted to the history of aviation of the First World War. There described in detail the course of military operations, the structure of the air forces of different countries, the biographies of pilots (especially aces), the history of various aviation units, as well as the material part and the history of the construction of warplanes of this period. Furthermore, a certain place is assigned to the colors and markings of the aircraft.

While working on the reference book *French aces of the First World War*, published in Russian in 2018, the authors of this album accomplished hundreds of color profiles of the planes on which the French aces of the First World War flew (including a small number that managed to fight during the Second World War). However, only 180 such profiles were included in the directory. So our task in this publication is to release about 500 color images of French aces' planes, many of which are published for the first time.

As a matter of course, the authors had to process an enormous aggregation of printed material such as books, magazines, and other periodicals over the many years. The basis for creating profiles was foremost photos, though some of them were made according to the narrative descriptions of eyewitnesses of the events of 1914–1918, although there were relatively few of them. A significant contribution was also made by Internet sites, where we managed to find a great deal of useful information. Finally, personal communication with serious historians on this topic has also greatly enriched this work. The book includes not only fighter planes on which aces gained the majority of their victories, but also military aircraft of other purposes, on which pilots began their combat careers, up to those on which they were trained in flying skills. Apart from the side images of aircraft, the book also includes some of their top-view projections when important marking elements were placed on the wings or on the top of the fuselage that could not be seen on the side projection. The book took the authors more than three years of painstaking work, and now this publication is the most complete collection of color profiles of French aces of the First World War, published in one edition.

We only hope that not only historians of the First World War and technology of that period will be interested in this work, but also modelers, manufacturers of models and aircraft decals in various scales, and a wide range of readers taking interest in various aspects of military and aviation history. So everyone could easily find a great deal of useful and really unique information.

Anton Zhukov and Alexander Kazakov

Albert Achard (1894–1972)

Capitaine
MS 48, Esc 91S, MF 98, F 203, N 510, N 85, Spa 78, Spa 150
5 victories

Nieuport 17
N 85
July–August 1917

SPAD 7
Spa 23
August 1917
Achard probably flew this Spad 7

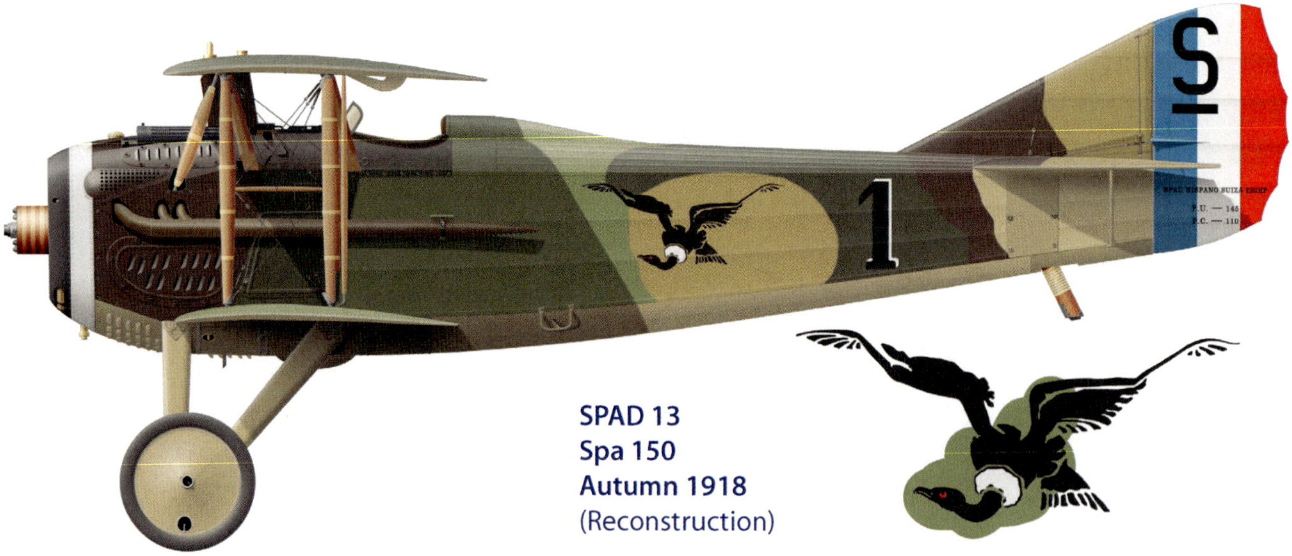

SPAD 13
Spa 150
Autumn 1918
(Reconstruction)

Maurice Jean Paul Etzeard "Marc" Ambrogi (1896–1971)

Sous-Lieutenant
N 507, N 90, Spa 90
14 victories (+1 in WW2)

Nieuport 23 No.N2782
N 90
August–September 1917

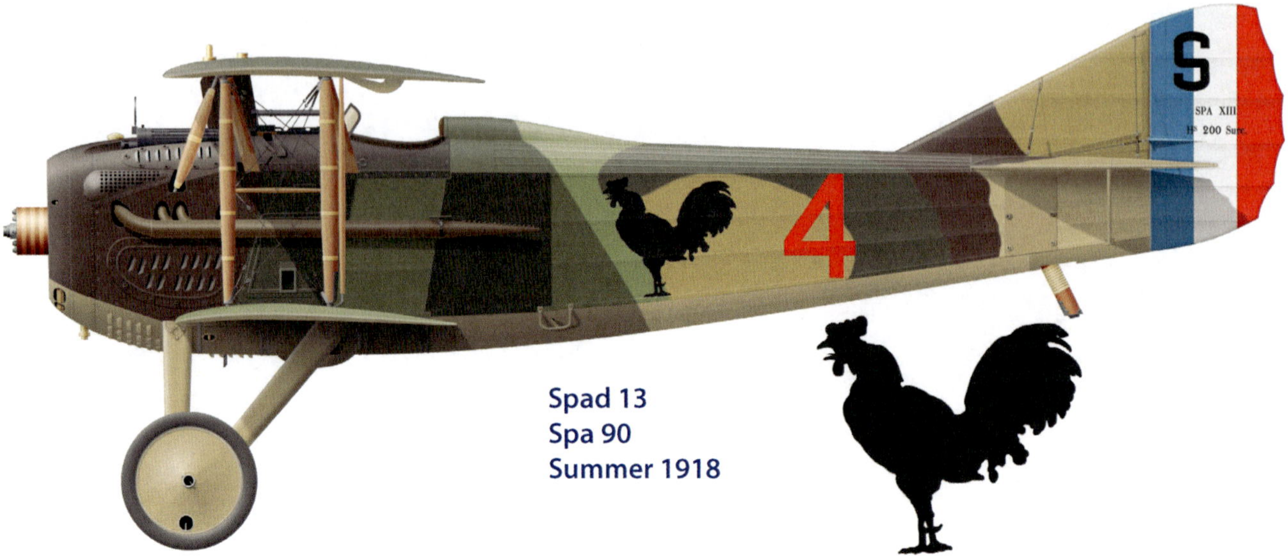

Spad 13
Spa 90
Summer 1918

Bloch MB.152 No.231 (Y-718)
GC 1/8
May 1940 r.

Argeev Pavel Vladimirovich (Paul d’Argueeff) (1897-1922)

Capitaine
N 48, 12th AOI, XIX KAO, Spa 124
15 victories

Nieuport 17 No.1908
XIX KAO
Spring 1917

Spad 13 No. S4872
Spa 124
July 1918

Maurice Albert Alfred Jean Arnoux
(1895–1940)
Sous-Lieutenant
MF99, MF55, N49, Spa49
5 victories

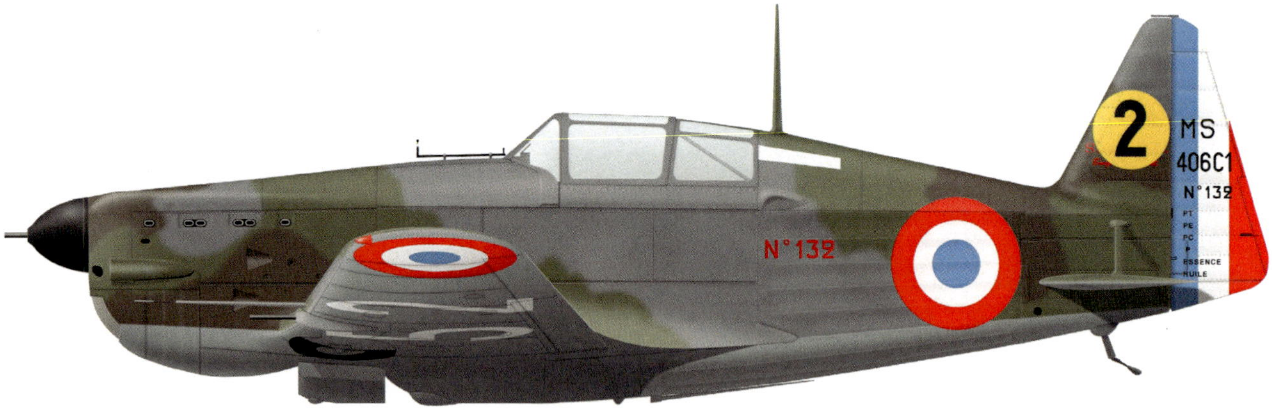

Morane-Saulnier MS 406C1 '2' No. 132
6eme Escadrille GC III/7
October 1939

Morane-Saulnier MS 406C1'6' No. 813
6eme Escadrille GC III/7
May 1940

Morane-Saulnier MS 406C1'2' No. 132
6eme Escadrille GC III/7
June 1940

Jean Henry Maurice Arpheuil
(1892–1920)
Lieutenant
Spa 151, Spa 170
5 victories

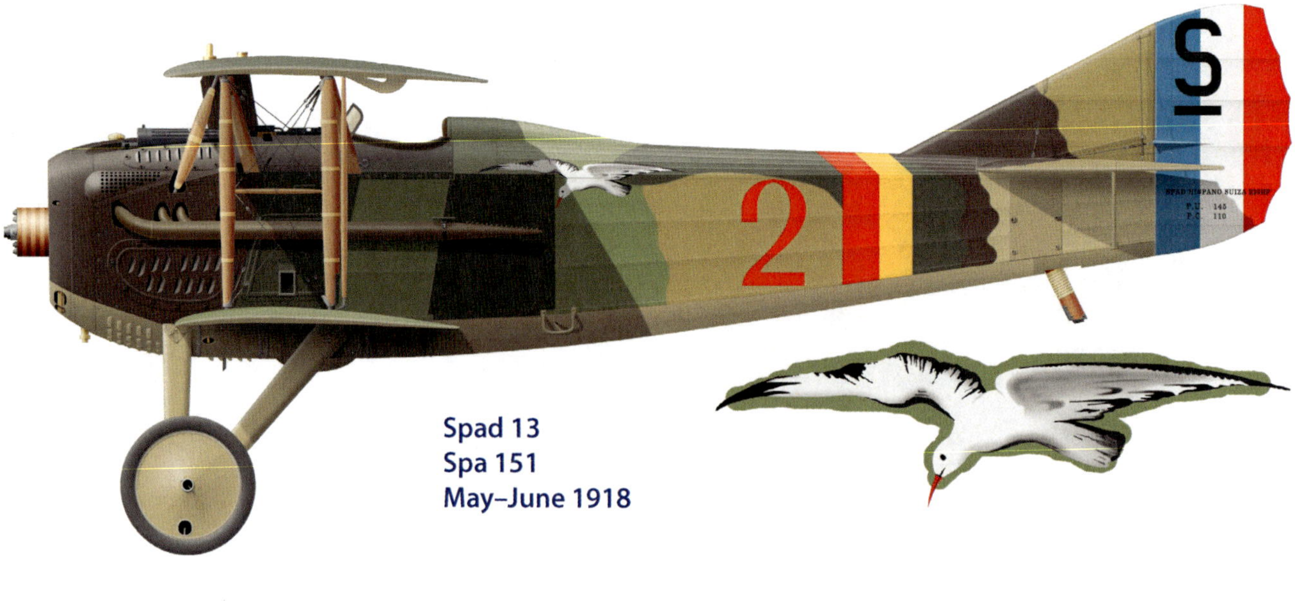

Spad 13
Spa 151
May–June 1918

Bernard Artigau
(1894–1968)
Sous-Lieutenant
N 15, Spa 15
12 victories

Spad 13 No. S8455
Spa 15
August 1918

Bernard Artigau

Artigau
Spad 13 No. S8455
Spa 15
August 1918

Spad 13
Spa 15
August 1918

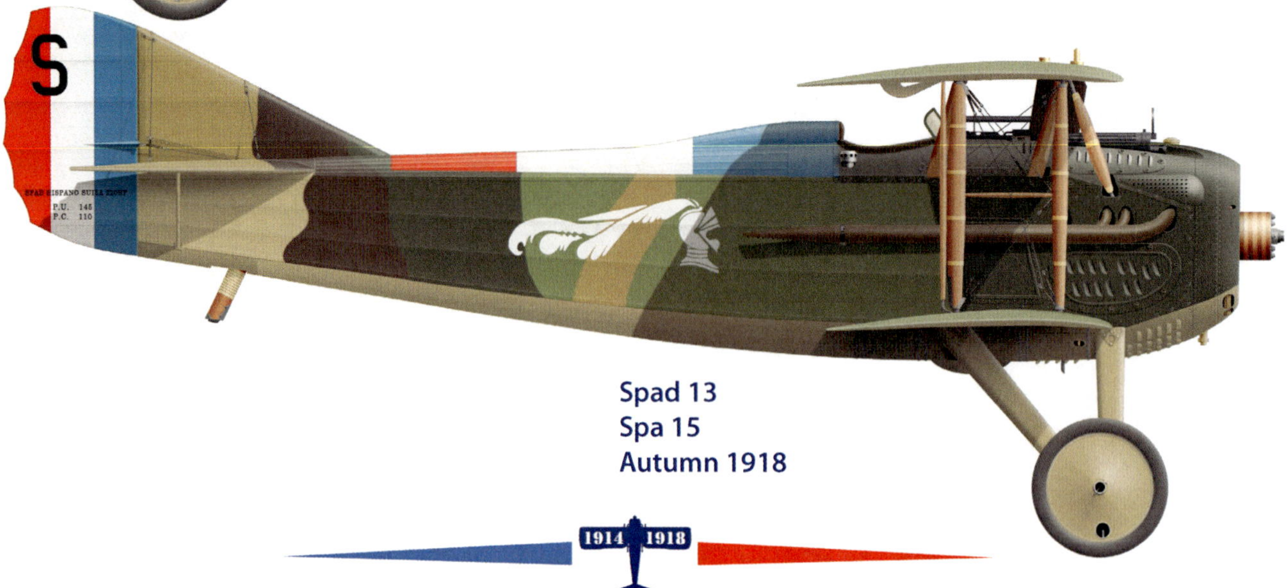

Spad 13
Spa 15
Autumn 1918

Alfred Victor Robert Auger
(1889–1917)
Capitaine
C 11, N 31, N 3
7 victories

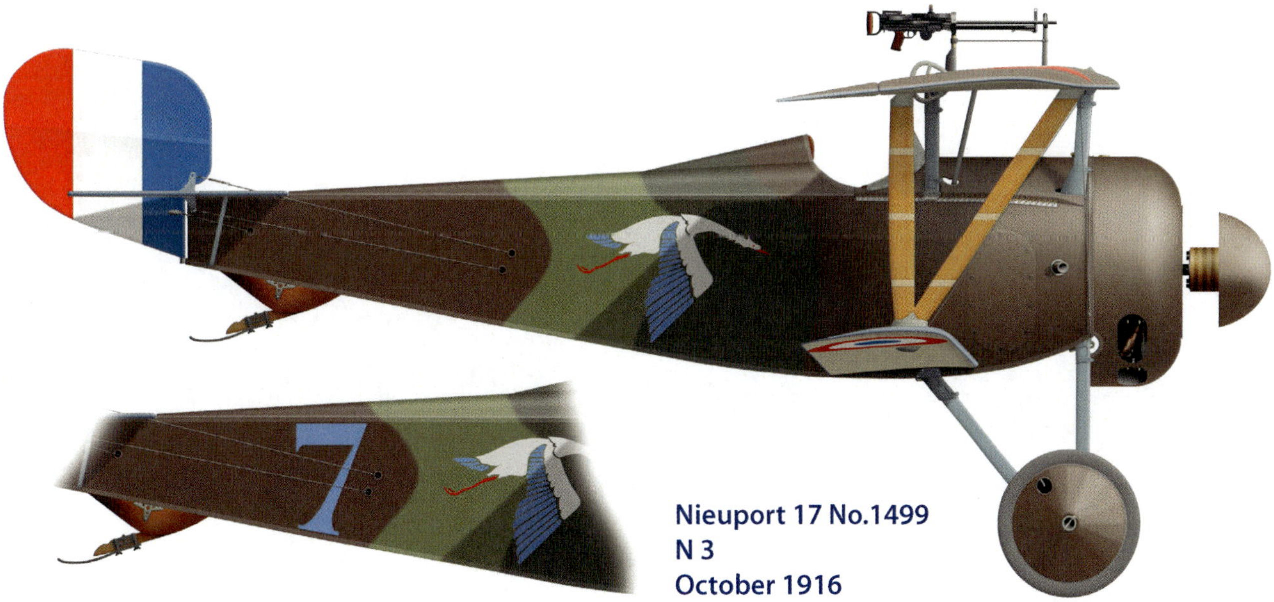

Nieuport 17 No.1499
N 3
October 1916

Alfred Auger

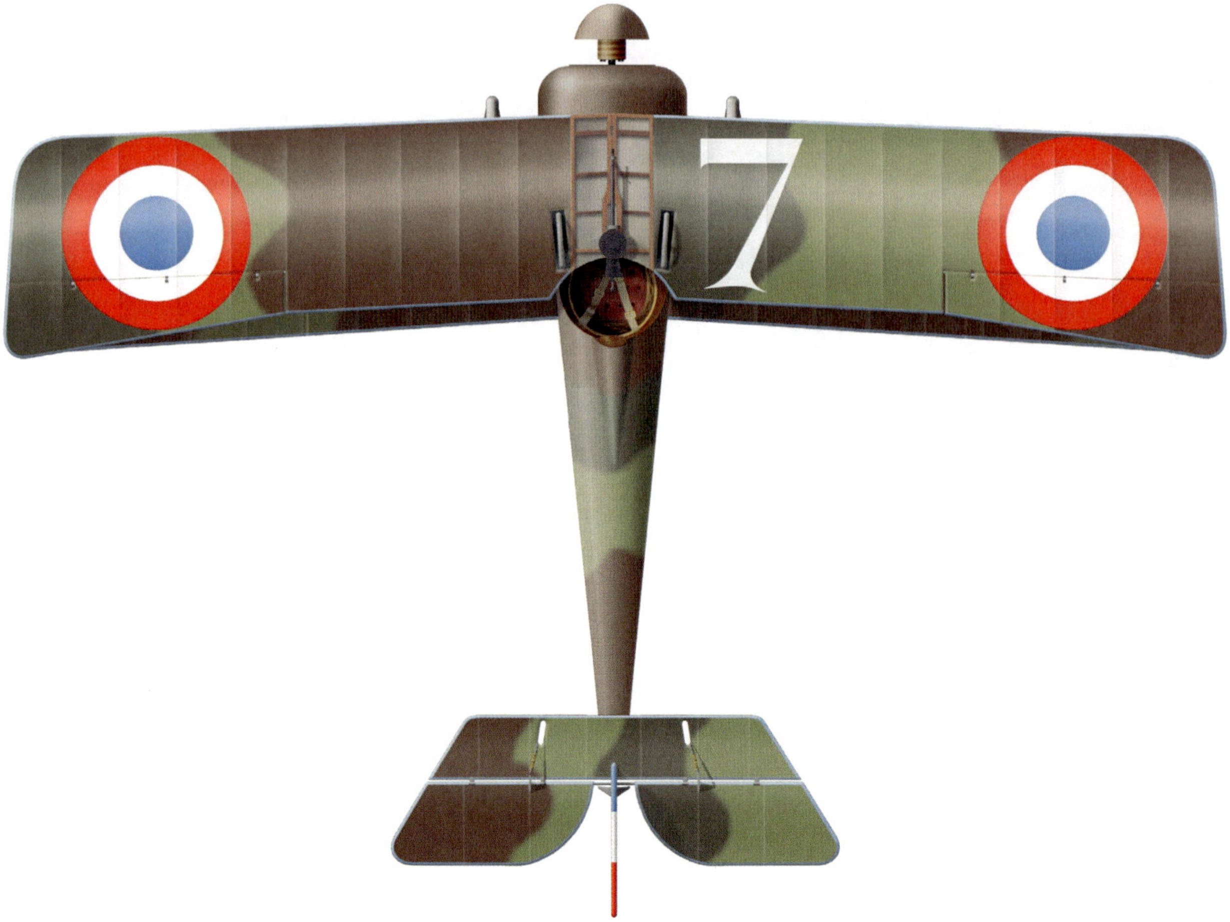

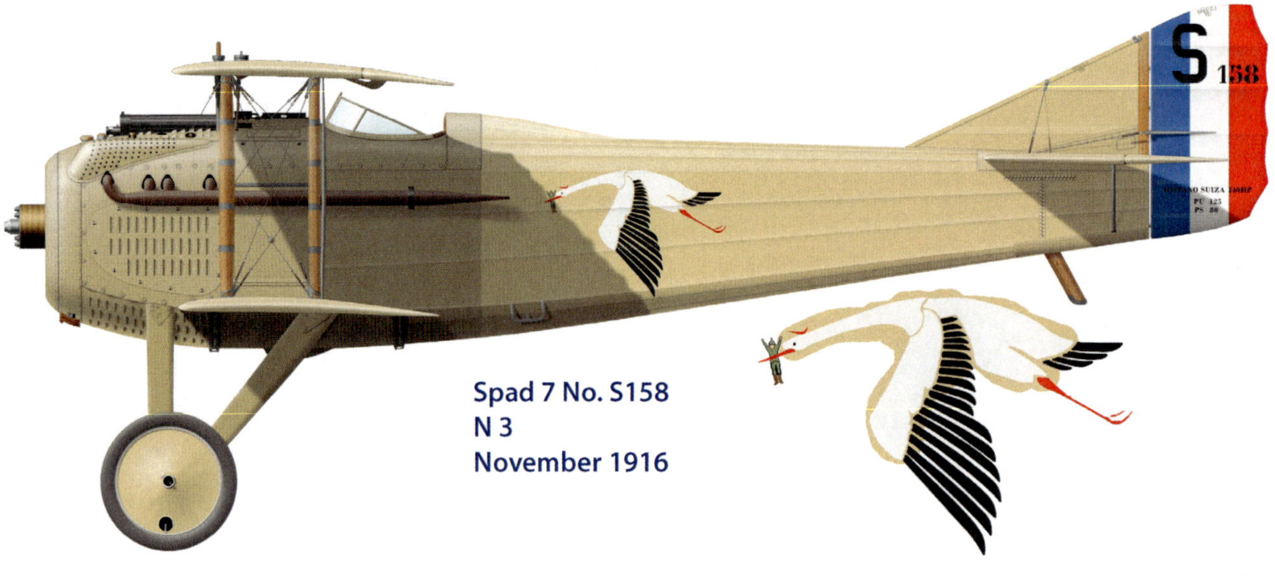

Spad 7 No. S158
N 3
November 1916

Spad 7 *Je* No. S1416
N 3
May 1917

Spad 7 *Je* No. S1416
N 3
June–July 1917

Spad 7
N 3
28 July 1917
(Probably No. S1729

Yves Felix Barbaza
(1883–1971)
Sous-Lieutenant
N 38, Spa 77
7 victories

Spad 13
Spa 77
Autumn 1917

Bernard Henri Barny de Romanet

(1894–1921)
Sous-Lieutenant
C 51, N 37, Spa 167
18 victories

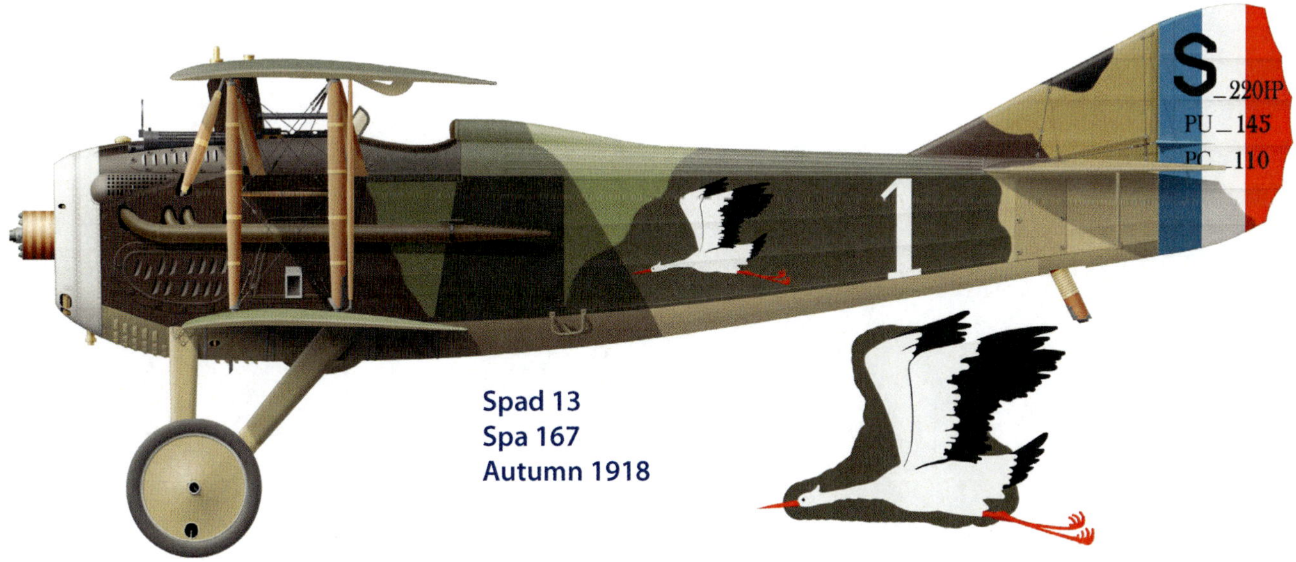

Spad 13
Spa 167
Autumn 1918

François Marie Noel Battesti
(1890–1977)
Sous-Lieutenant
BL 18, BL 3, C 10, N 73, Spa 73
7 victories

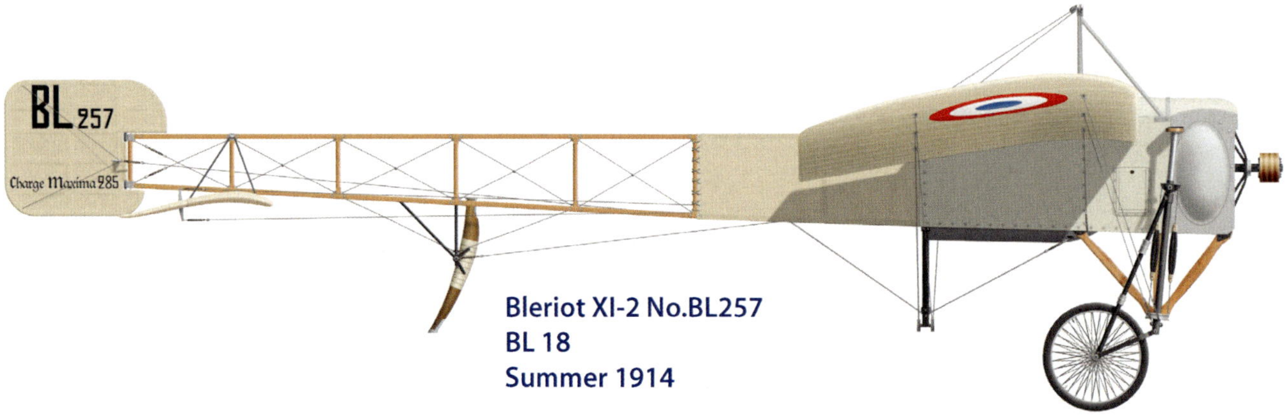

Bleriot XI-2 No.BL257
BL 18
Summer 1914

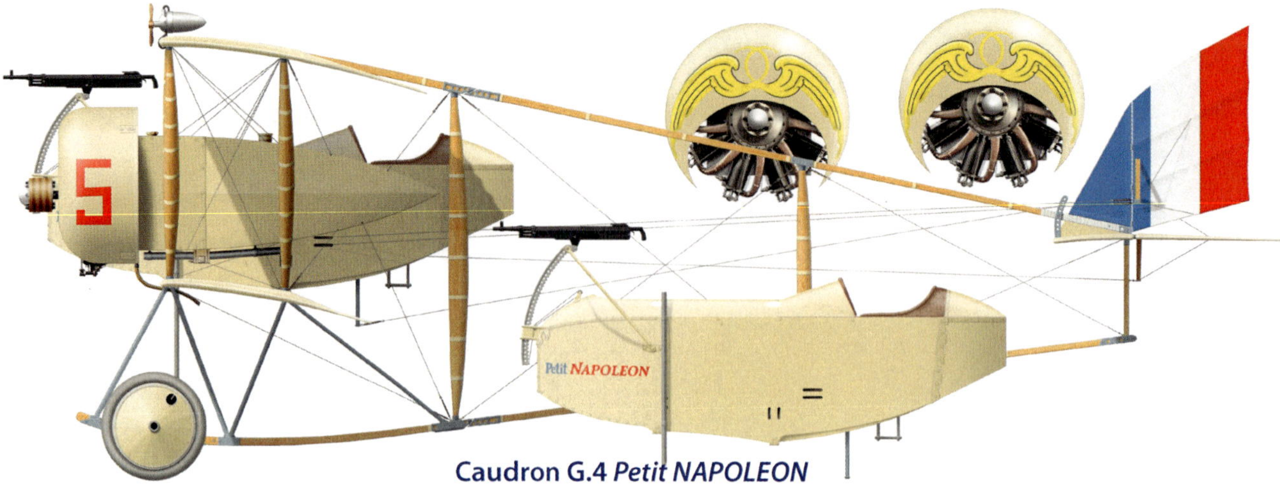

Caudron G.4 *Petit NAPOLEON*
C 10
Summer 1916

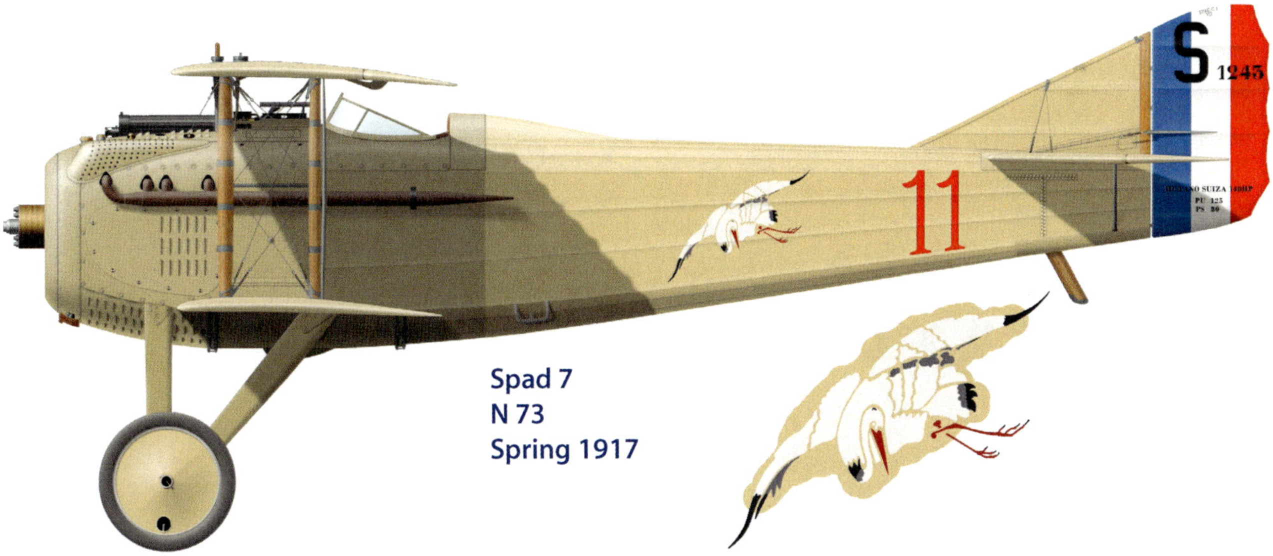

Spad 7
N 73
Spring 1917

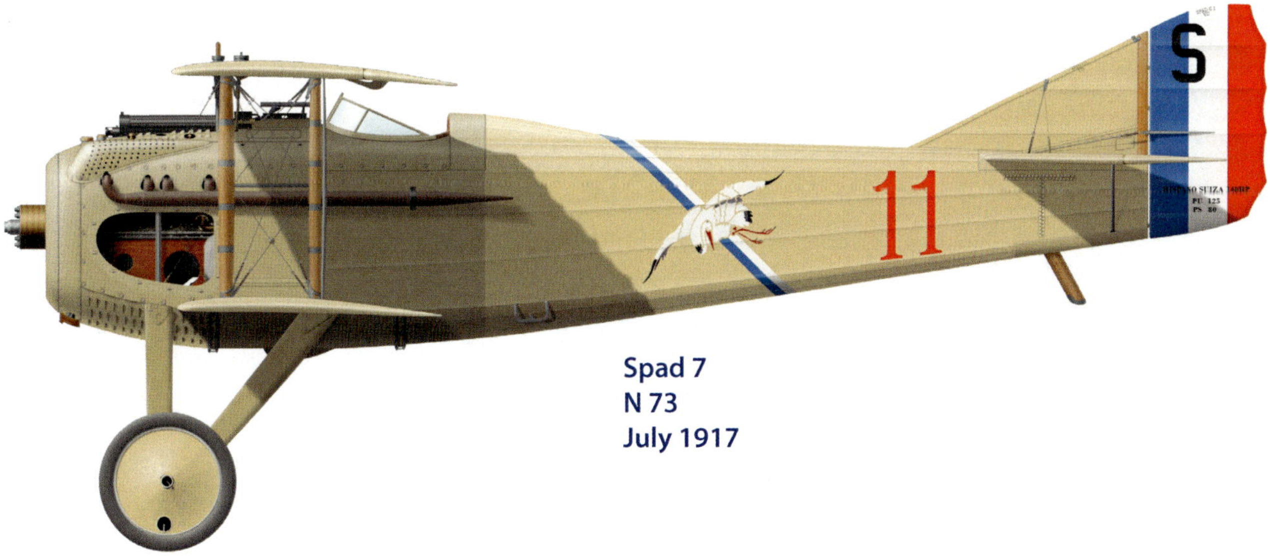

Spad 7
N 73
July 1917

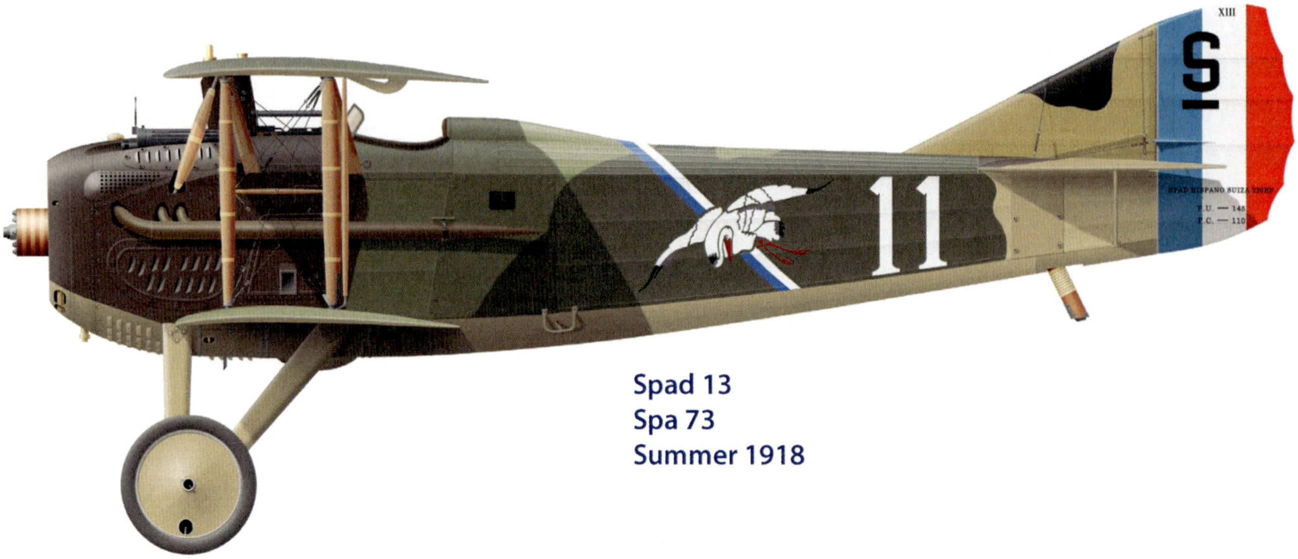

Spad 13
Spa 73
Summer 1918

Spad 17 No.S694
Spa 73
October 1918

Frank Leaman Baylies
(1895–1918)
Lieutenant
Spa 3, Spa 73
12 victories

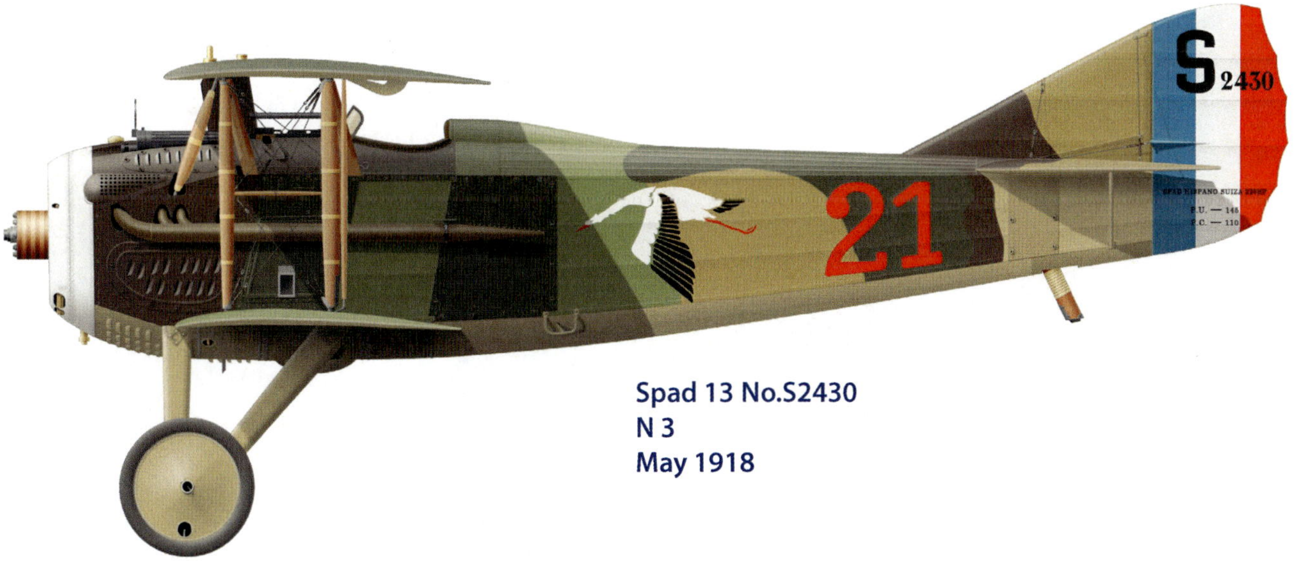

Spad 13 No.S2430
N 3
May 1918

Armond Jean Berthelot

(1894–1961)
Adjudant
Esc 365, N 15, Spa 15
11 victories

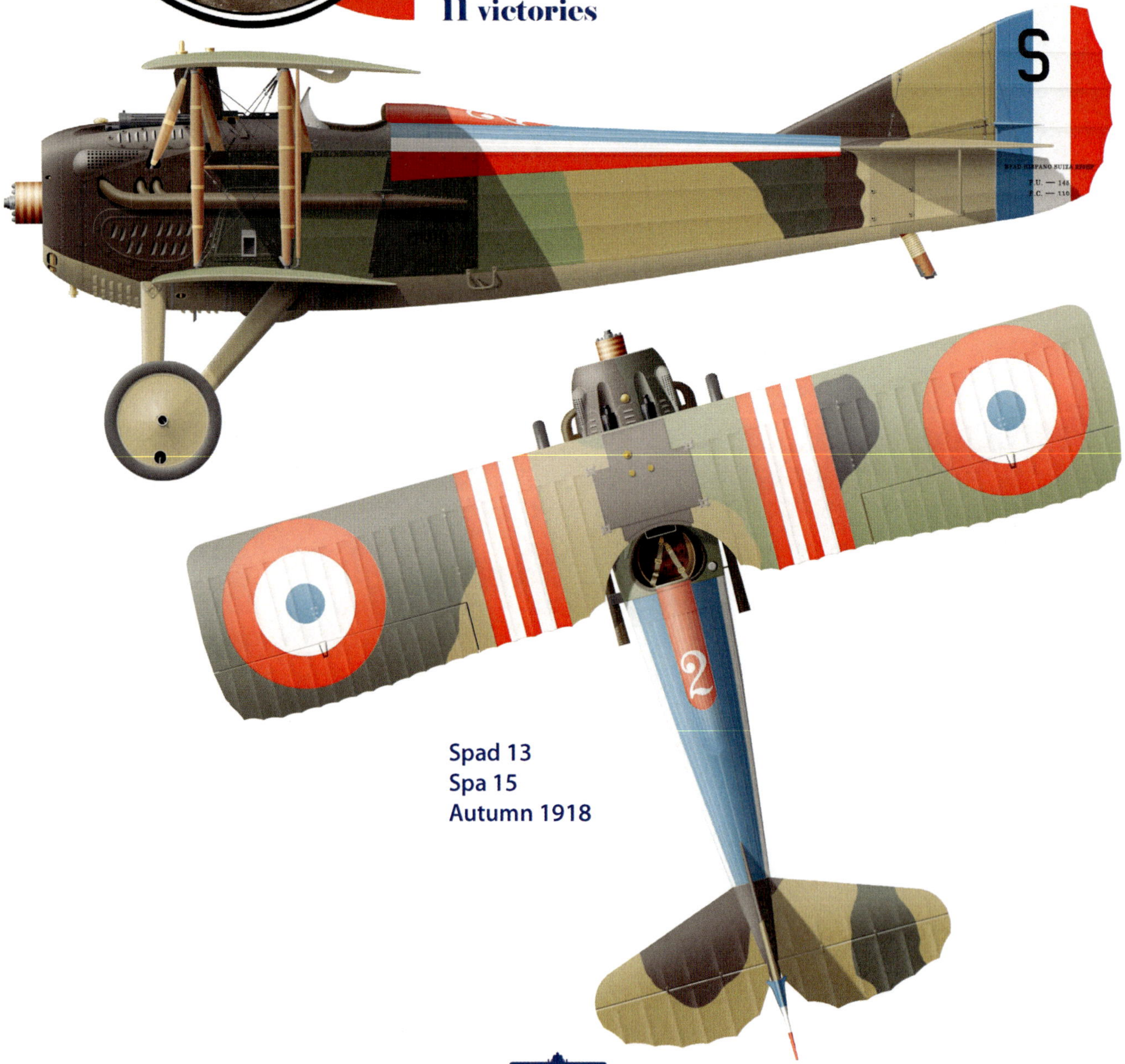

Spad 13
Spa 15
Autumn 1918

Pierre-George Blanc
(1887–1960)
Adjudant
C 47, N 31, Spa 154
6 victories

Spad 7 No. S3240
N 31
August–September 1917

Marcel Robert Léopold Bloch
(1890–1938)
Sous-Lieutenant
N 3, N 62, Spa 62
5 victories

Spad 7 *Fantasio II*
Spa 62
Autumn 1917

François Eugène Marie Antoine de Boigne
(1899–1970)
Sous-Lieutenant
N 82, Spa 69
6 victories

Spad 7 No. S1405
Spa 82
June 1917

Adolphe, Alois de Gonzague, Marie Hubert Ghislain du Bois d'Aische (1875–1958)

Sergent
F 71, AR 71
6 victories

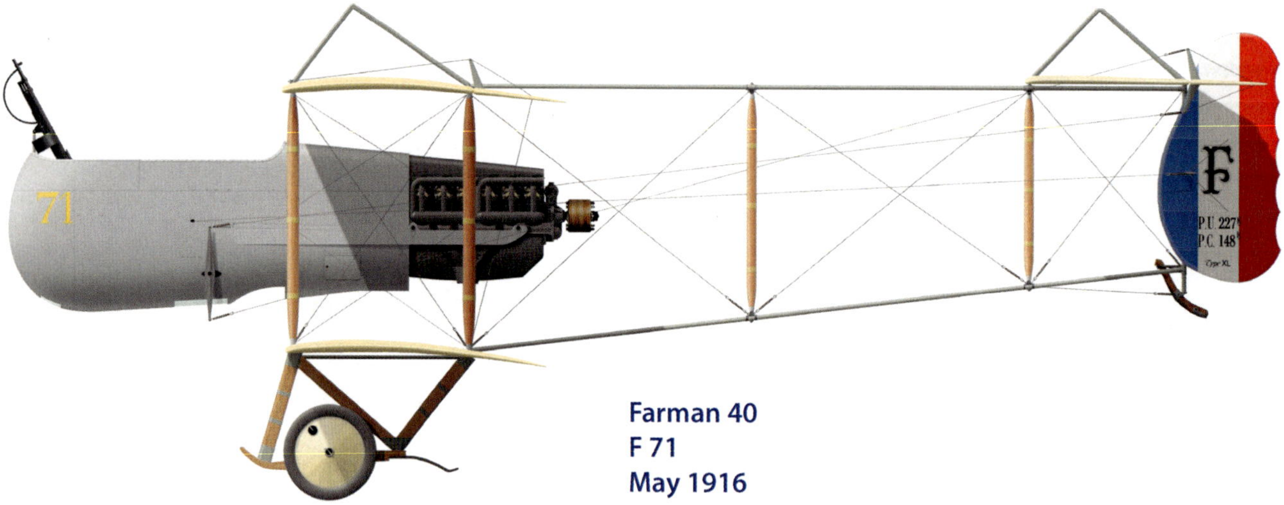

Farman 40
F 71
May 1916

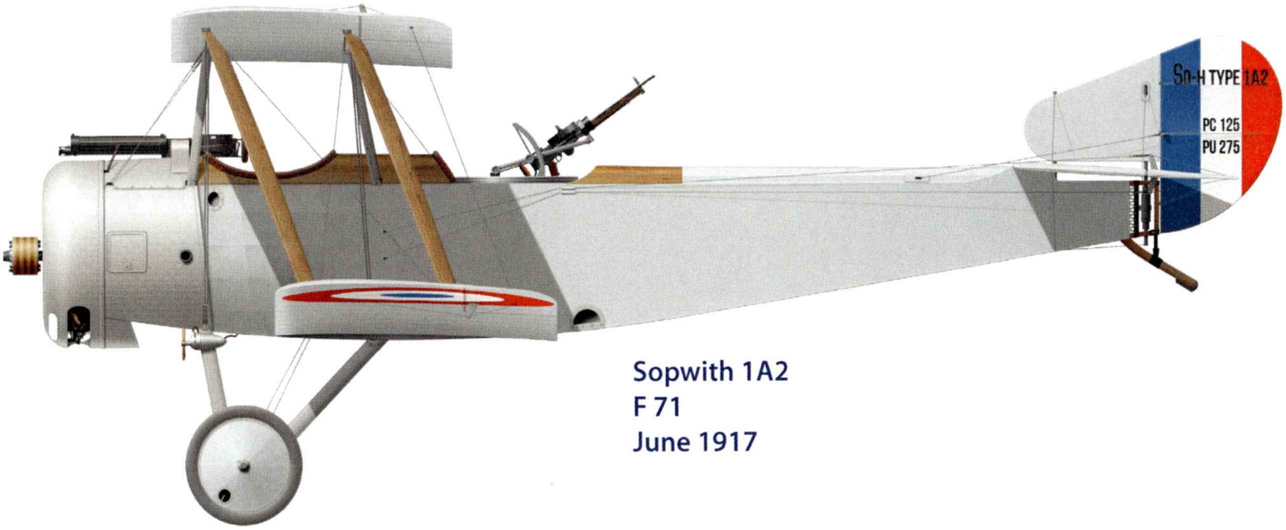

Sopwith 1A2
F 71
June 1917

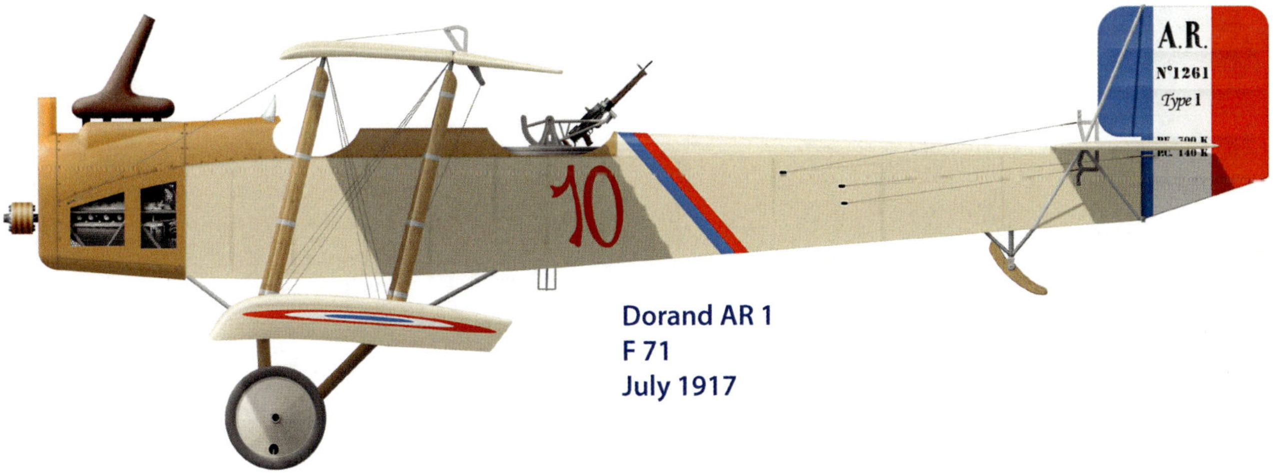

Dorand AR 1
F 71
July 1917

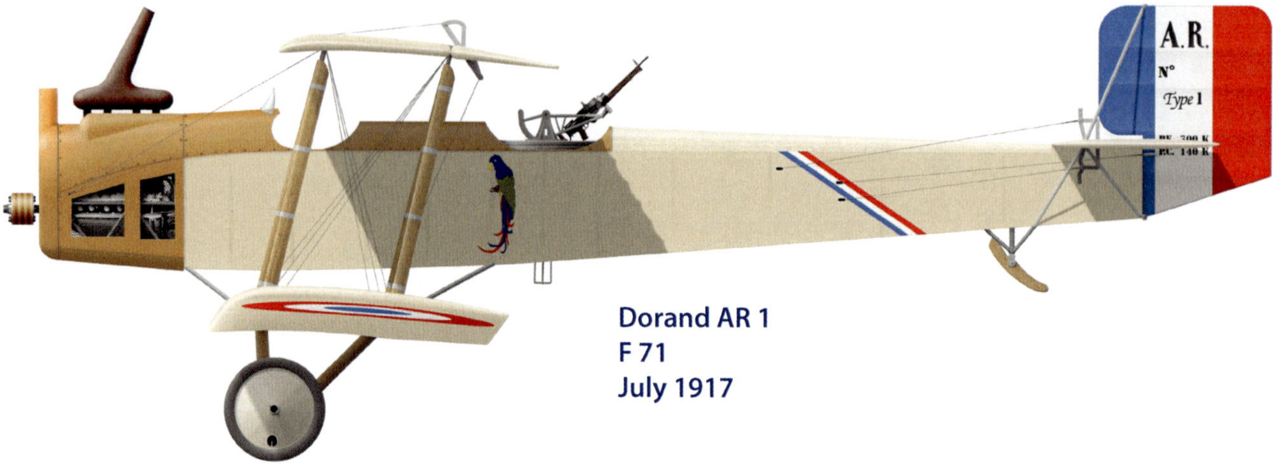

Dorand AR 1
F 71
July 1917

Letord 1
AR 71
August–September 1917

Honoré Marie Joseph Léon Guillaume de Bonald (1894–1977)

Lieutenant
N 69, Spa 69
5 victories

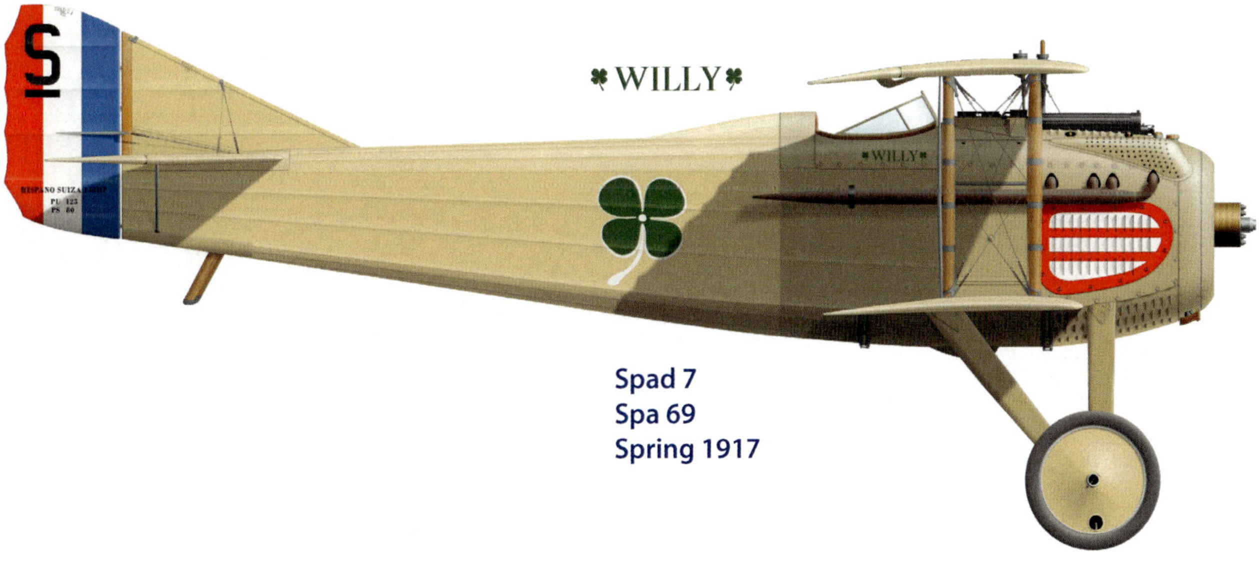

Spad 7
Spa 69
Spring 1917

Joseph Denis Bertrand Robert de Bonnefoy
(1894–1946)
Sous-Lieutenant
VB101 Br31, N65, N68, N84, Spa15, Spa23
6 victories

Spad 7
Spa 65
November 1916
(Reconstruction)

Fernand Régis Bonneton
(1890–1922)
Capitaine
Spa 69
9 victories

Spad 7
Spa 69
July–September 1918

Spad 13 No. S8305
Spa 69
May 1919

Charles Alexandre Bronislas Borzecki
(1880–1959)
Lieutenant
C 43, N 62, Spa 62
5 victories

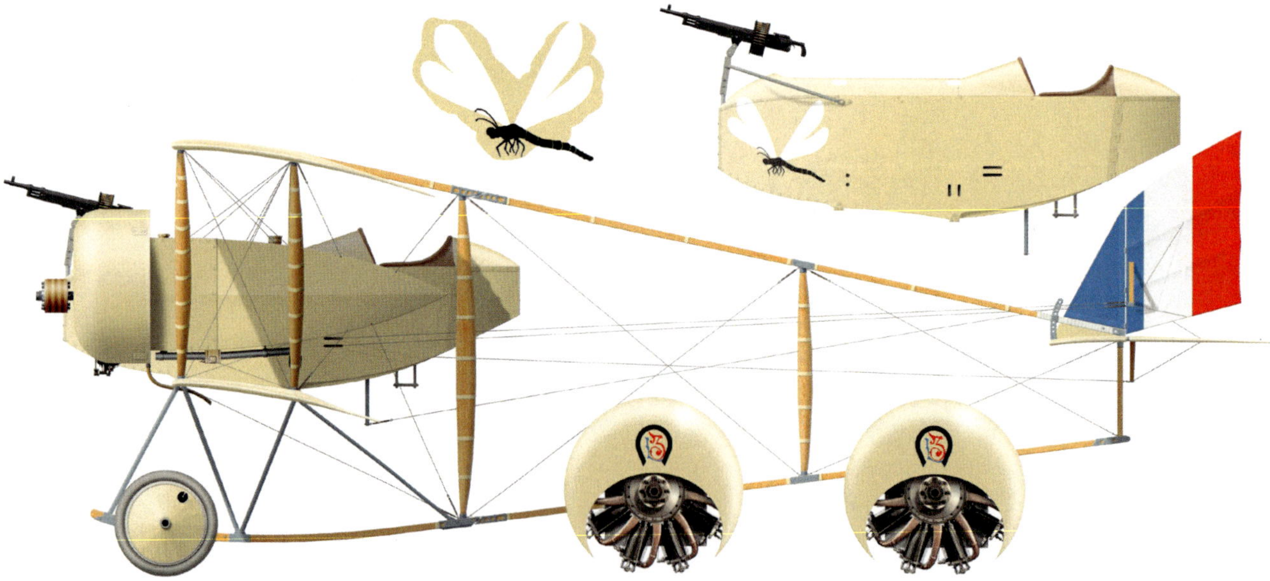

Caudron G.4 "L3"
C 43
July 1916

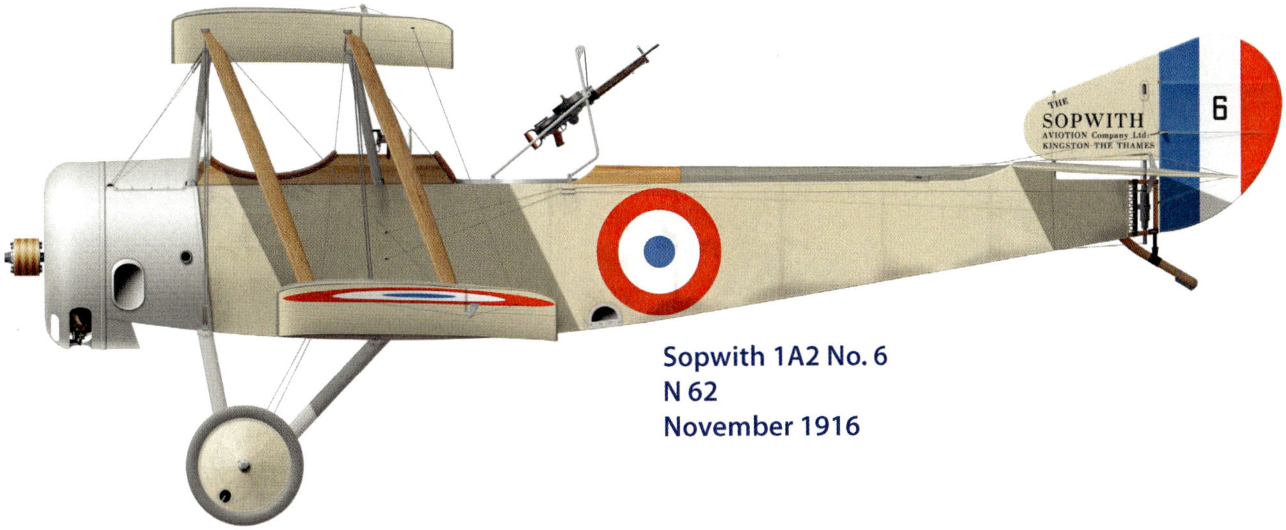

Sopwith 1A2 No. 6
N 62
November 1916

Sopwith 1A2 No. 6
N 62
Spring 1917

Spad 11
Spa 62
November 1917

André Louis Bosson

(1894–1918)
Sergent
Spa 62
7 victories

Spad 7 *Nenette* No. S5325
Spa 62
July 1918

Léon Jean-Pierre Bourjade
(1889–1924)
Sous-Lieutenant
N 152, Spa 152
28 victories

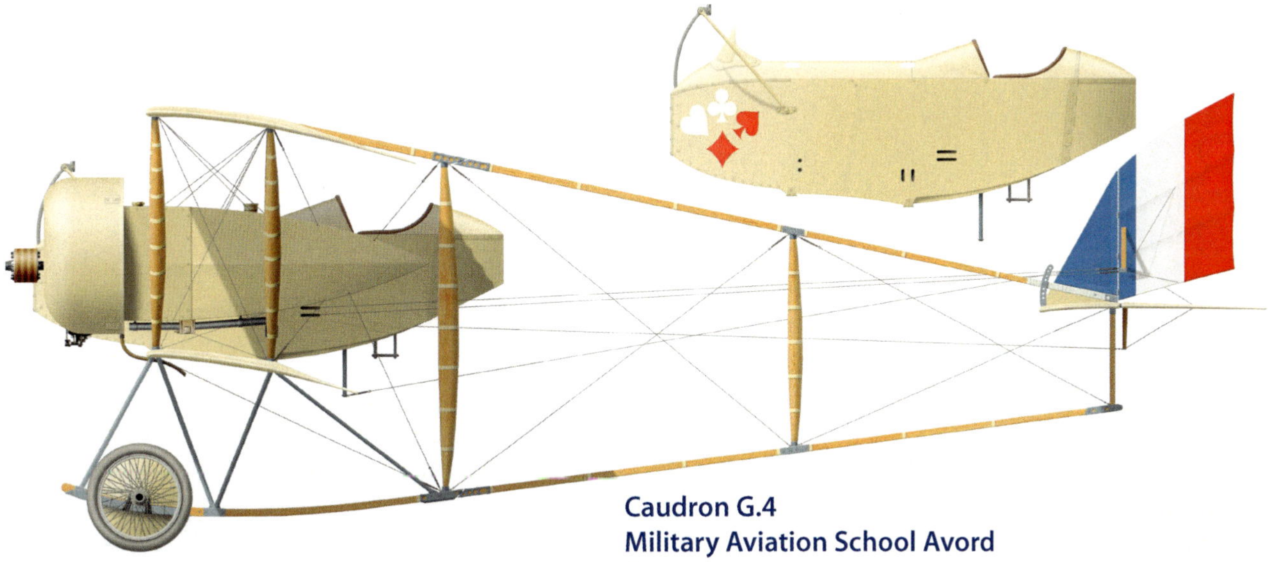

Caudron G.4
Military Aviation School Avord
June 1917

Nieuport 27
Spa 152
February 1918

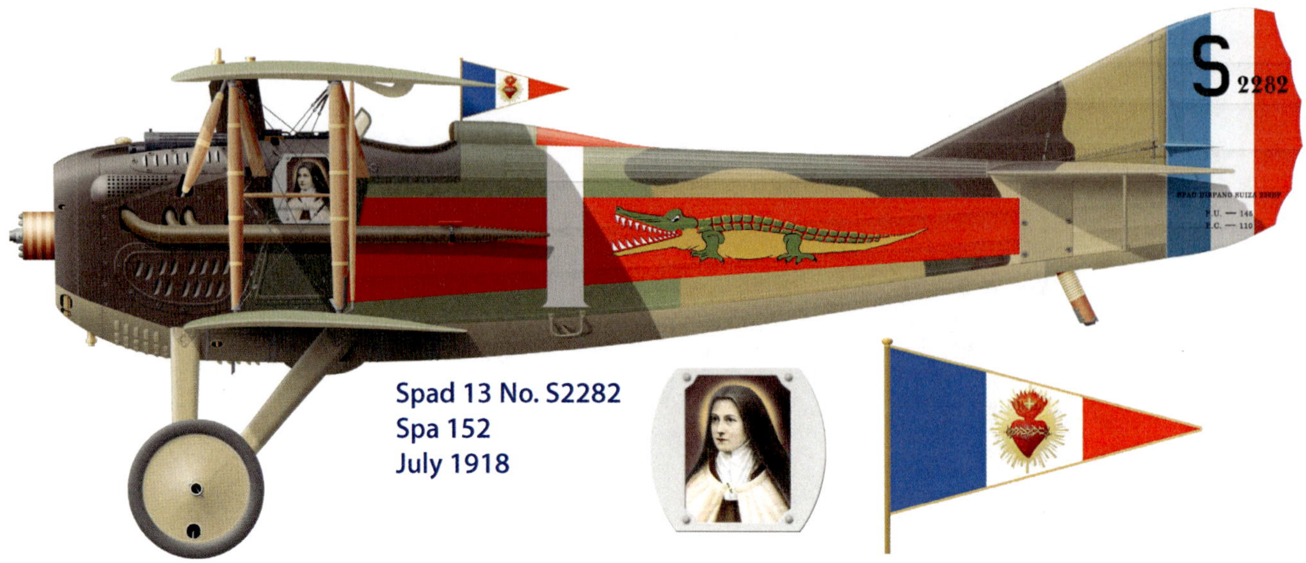

Spad 13 No. S2282
Spa 152
July 1918

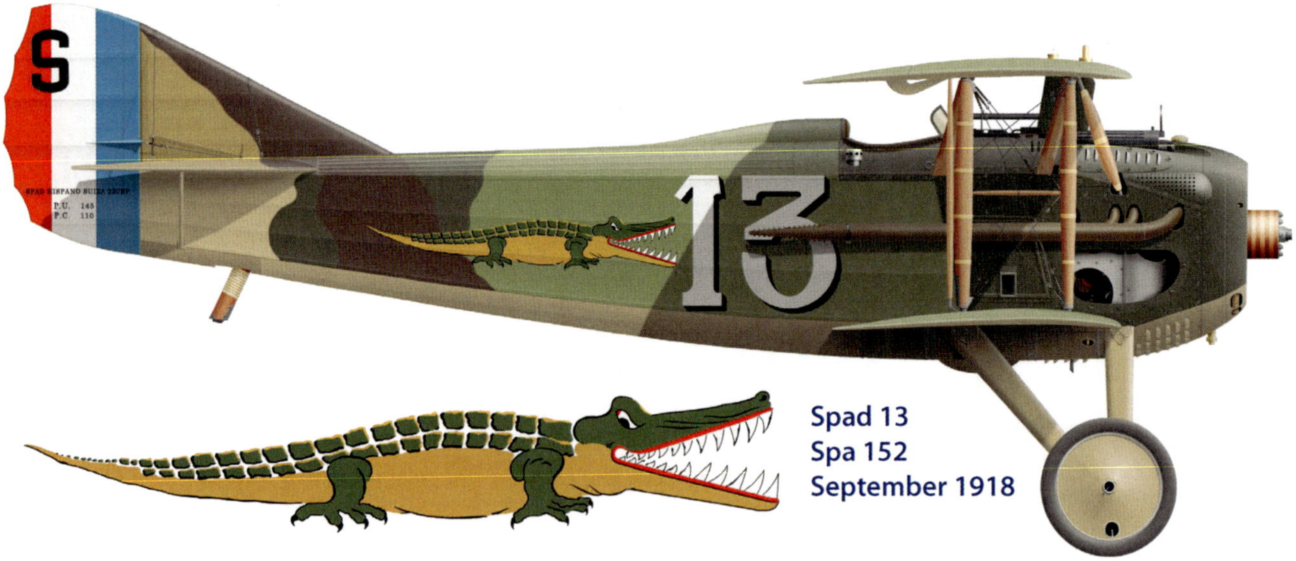

Spad 13
Spa 152
September 1918

Jean Georges Bouyer
(1891–1921)
Sous-Lieutenant
N 49, Spa 49
11 victories

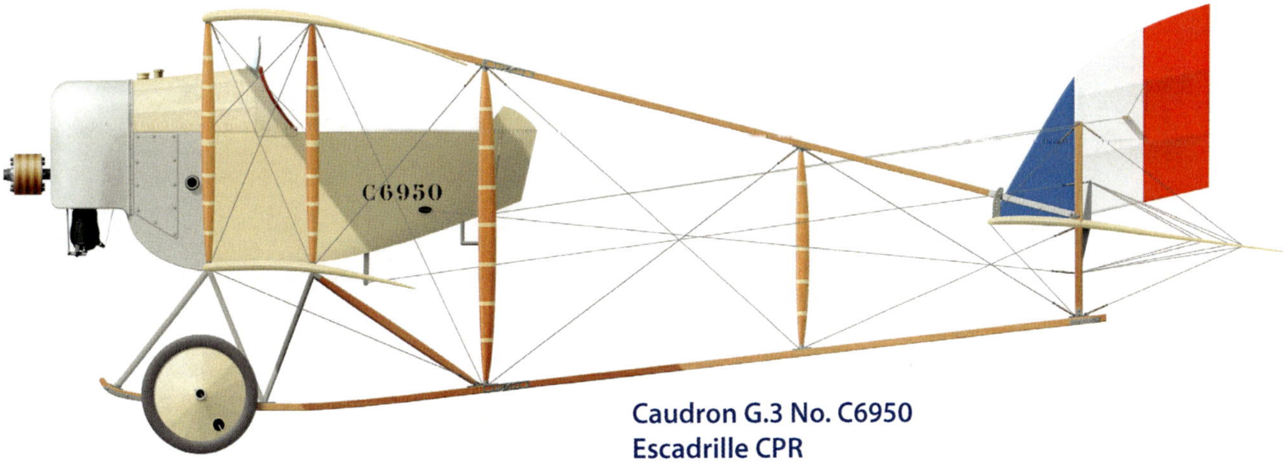

Caudron G.3 No. C6950
Escadrille CPR
Spring 1915

Maurice Jean Paul Boyau
(1888–1918)
Sous-Lieutenant
N 77, Spa 77
35 victories

Nieuport 17
N 77
Summer 1917

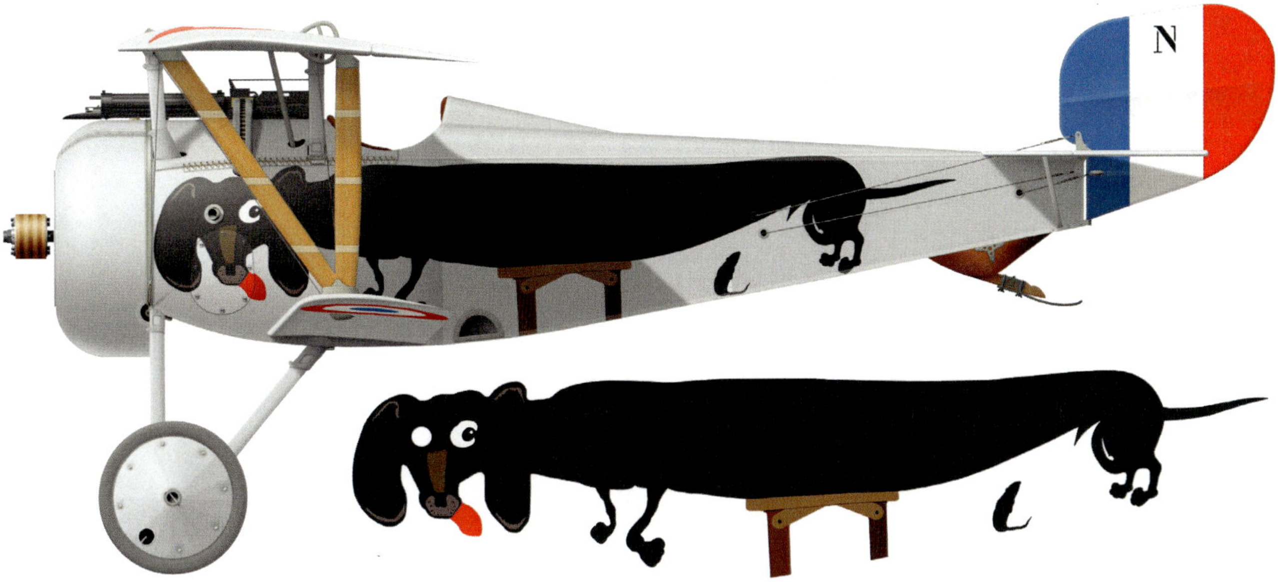

Nieuport 23
N 77
Autumn 1917

Nieuport 23
N 77
September 1917

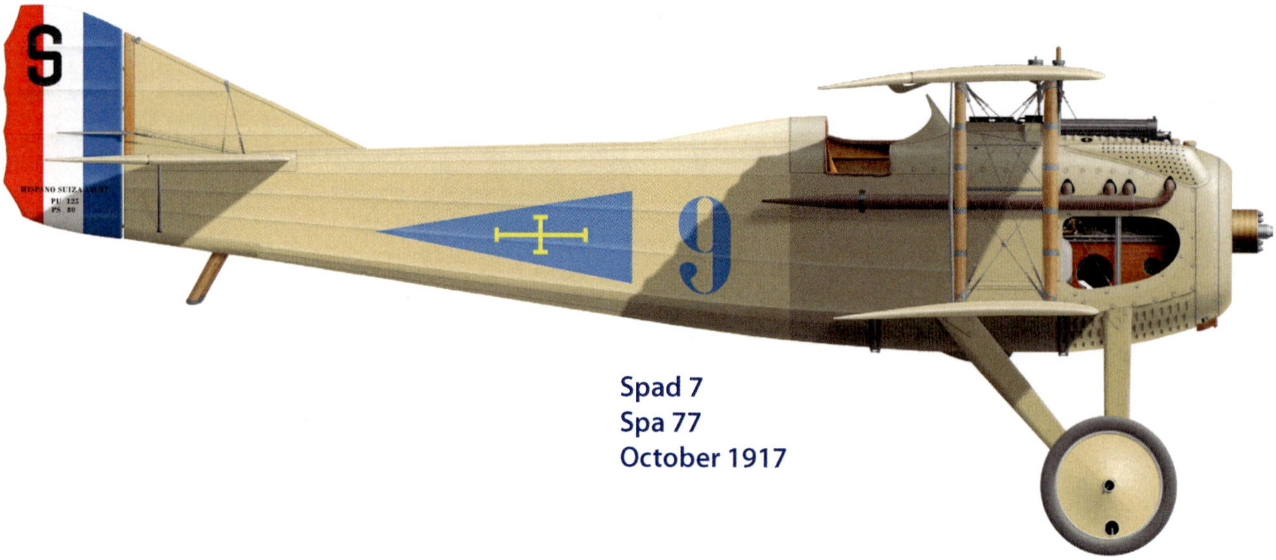

Spad 7
Spa 77
October 1917

Spad 13
Spa 77
Summer 1918

Jean Seraphin Benjamin Emmanuel Bozon-Verduraz (1889–1942)
Sous-Lieutenant
C 11, N 3, Spa 94
11 victories

Spad 13 *Mon Lion* No. S2533
Spa 3
May 1918

Alesandre Albert Roger Bretillon
(1895–1978)
Sous-Lieutenant
Spa 38, N 79, Spa 49
9 victories

Nieuport 17 No. 2454
N 79
August 1917

Spad 7
N 79
October 1917

Spad 7
N 79
February 1918

Alexandre Buisson
(1886–1939)
Adjudant
F 29, Sop 123, C 21, R240
5 victories

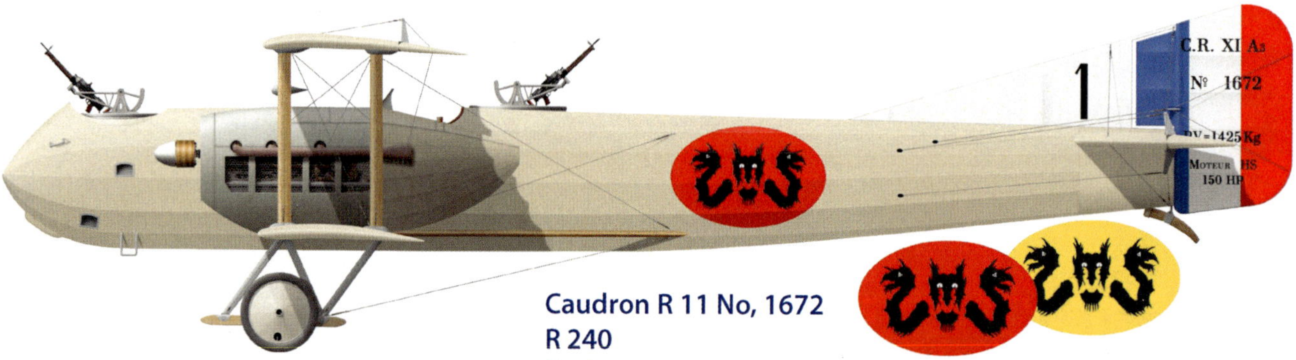

Caudron R 11 No, 1672
R 240
Spring 1918

Caudron R 11 No, 6514
R 240
October 1918

Edmond Eugene Henri Caillaux
(1896–1943)
Adjudant
MF88, Spa48
5 victories

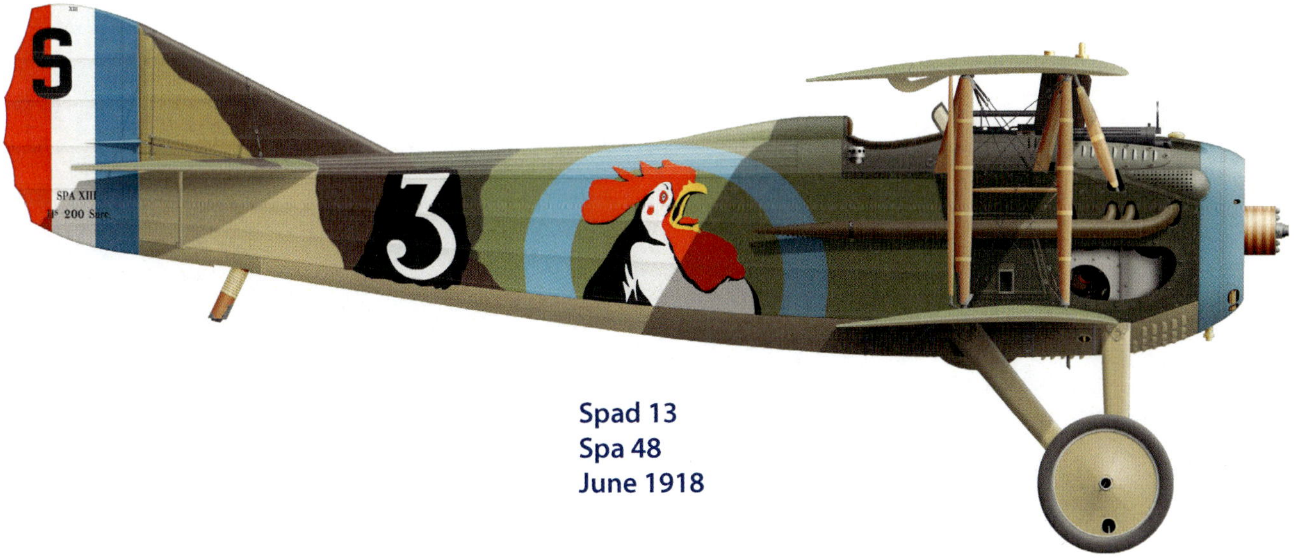

Spad 13
Spa 48
June 1918

Pierre Marie Joseph Cardon
(1894–1987)
Marechal-des-Logis
C 64, Spa 81
5 victories

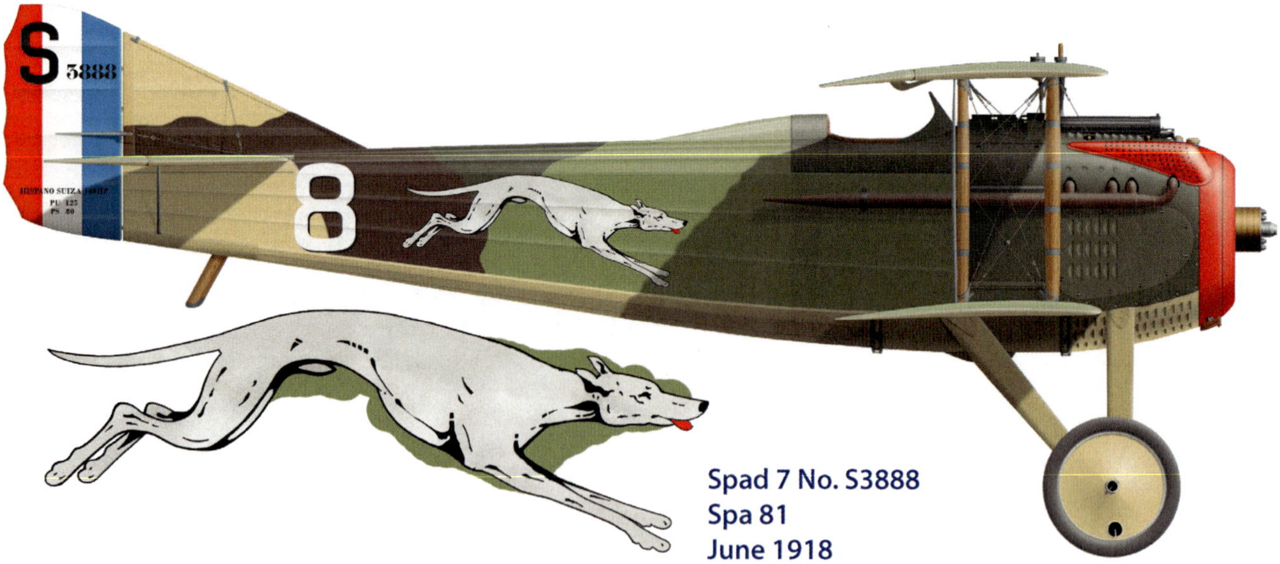

Spad 7 No. S3888
Spa 81
June 1918

Jean Pie Hyacinthe Paul Jérome Casale Marquis de Montferato (1893–1923)

Sous-Lieutenant

MF 8, N 23, N 67, Spa 38, Spa 156

13 victories

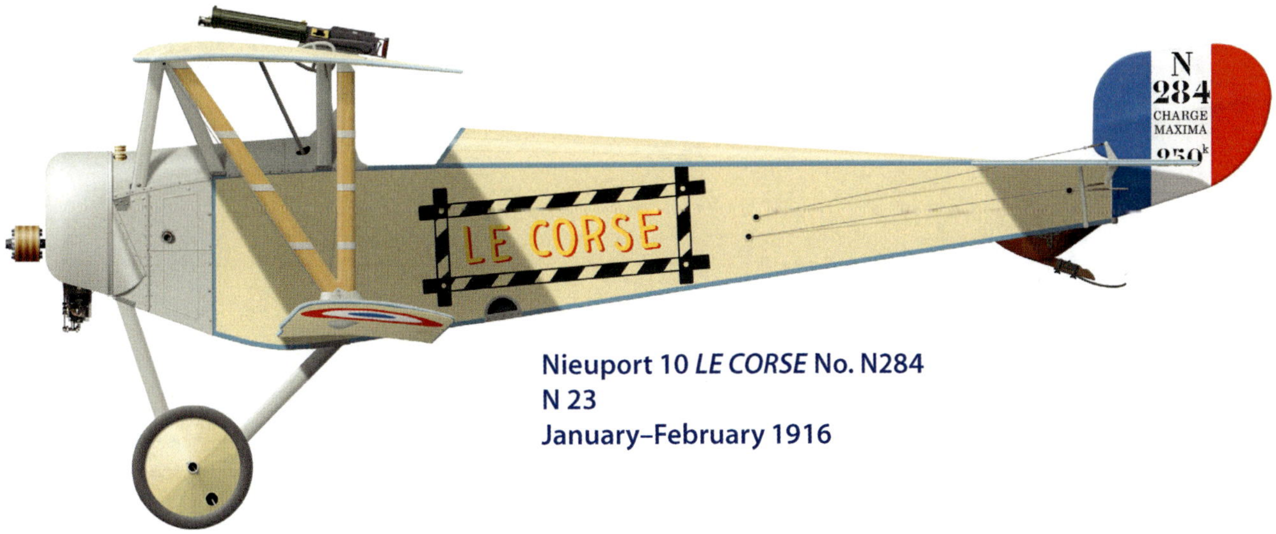

Nieuport 10 *LE CORSE* No. N284
N 23
January–February 1916

Nieuport 23 *Aj-t Lenoir*
N 23
Autumn 1916

Spad 7 *Vendetta Corsa*
N 23
Spring 1917

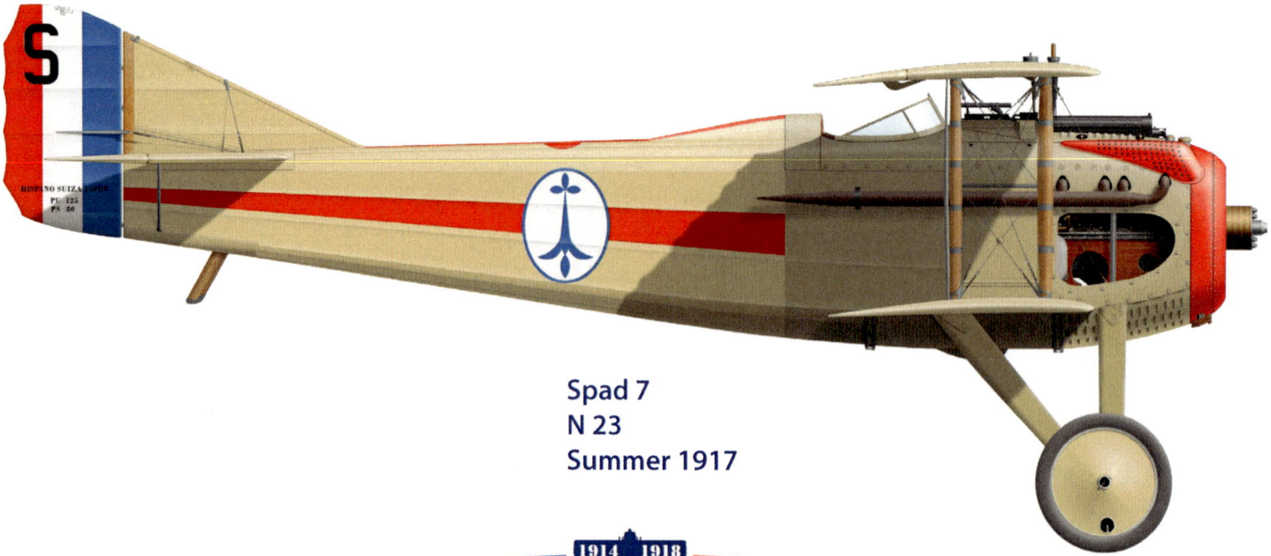

Spad 7
N 23
Summer 1917

Thomas Gantz Cassady

(1896–1972)
Capitan
Spa157, Spa163 (FAS), 28th Aero, 103rd Aero (USAS)
9 victories

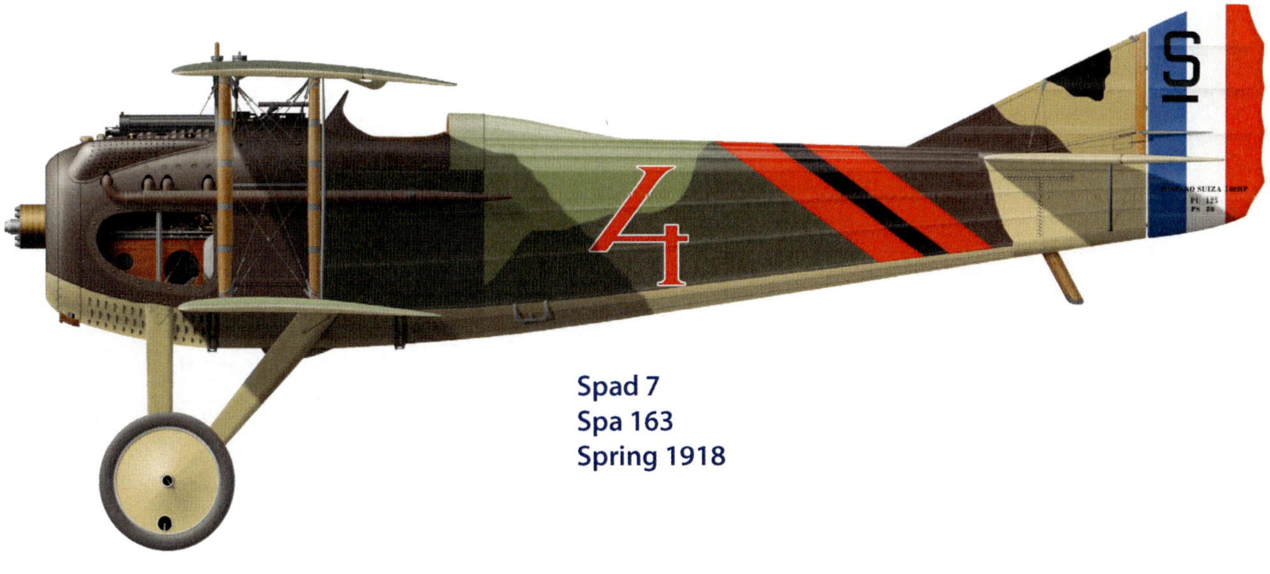

Spad 7
Spa 163
Spring 1918

Pierre Fortaner Paul de Cazenove de Pradines
(1894–1982)
Sous-Lieutenant
N 81, Spa 81
7 victories

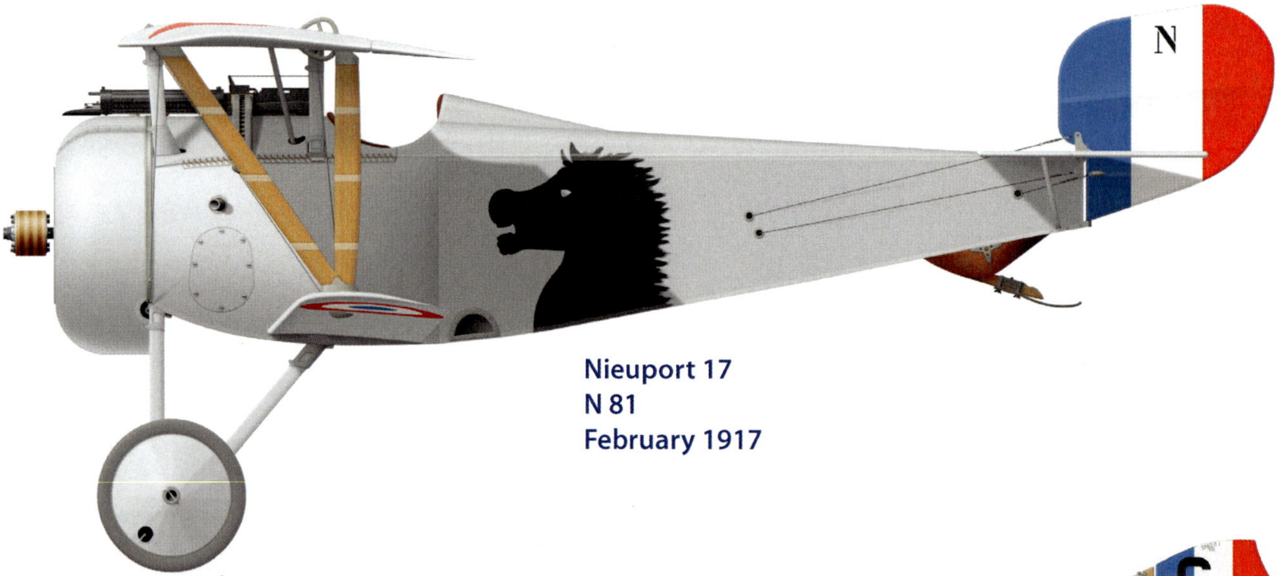

Nieuport 17
N 81
February 1917

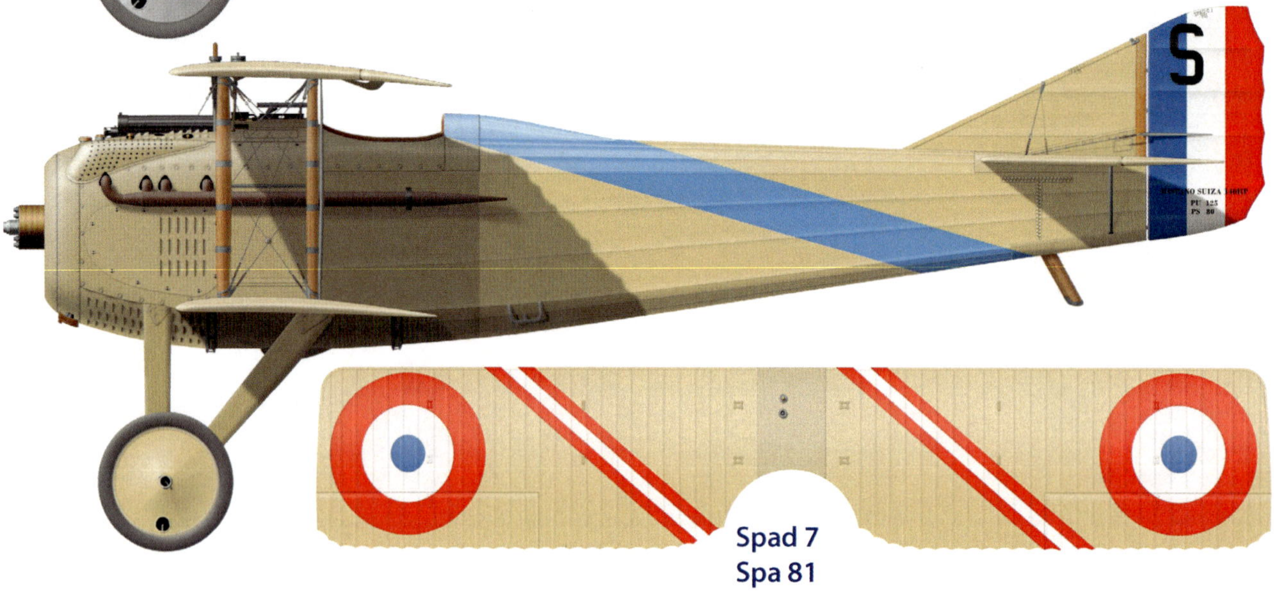

Spad 7
Spa 81
May–December 1917

Albert René Chabrier
(1896–1920)
Lieutenant
N 315
6 victories

Nieuport 27 No. N5276
Spa 315
Early 1918

Spad 7
Spa 315
October 1918

André Julien Chainat
(1892–1961)
Adjudant
BLC 5, MF 63, MS 23, MS 38, N 3, Spa 3
11 victories

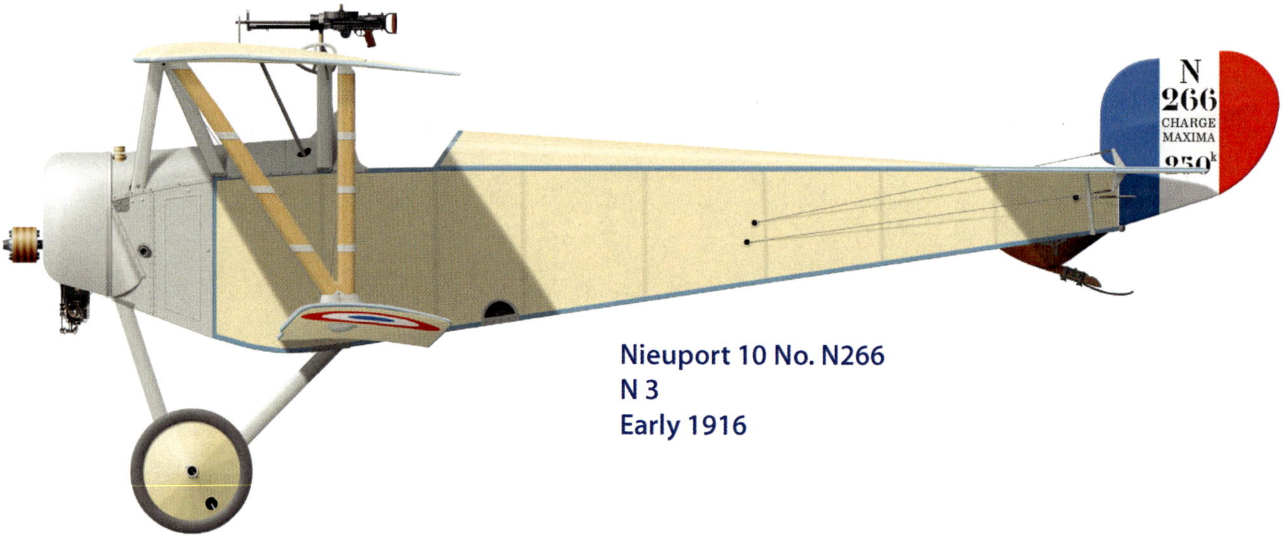

Nieuport 10 No. N266
N 3
Early 1916

Nieuport 11 No. N871
N 3
March 1916

Nieuport 16 No. N8952
N 3
April 1916

Nieuport 16 *l'Oiseau bleu 3* No. N1209
N 3
July 1916

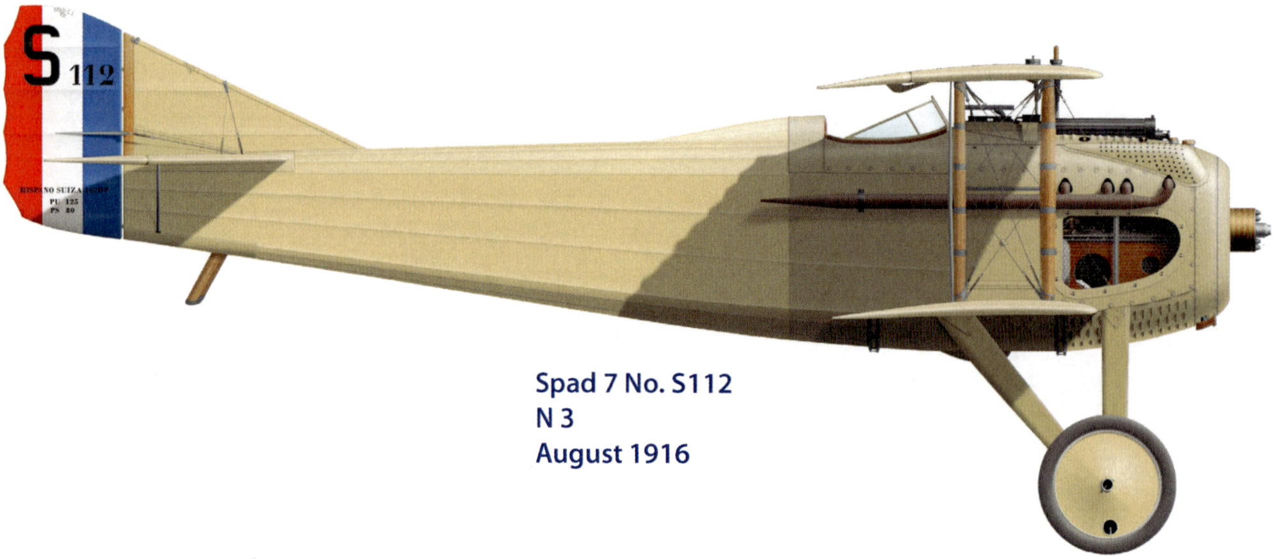

Spad 7 No. S112
N 3
August 1916

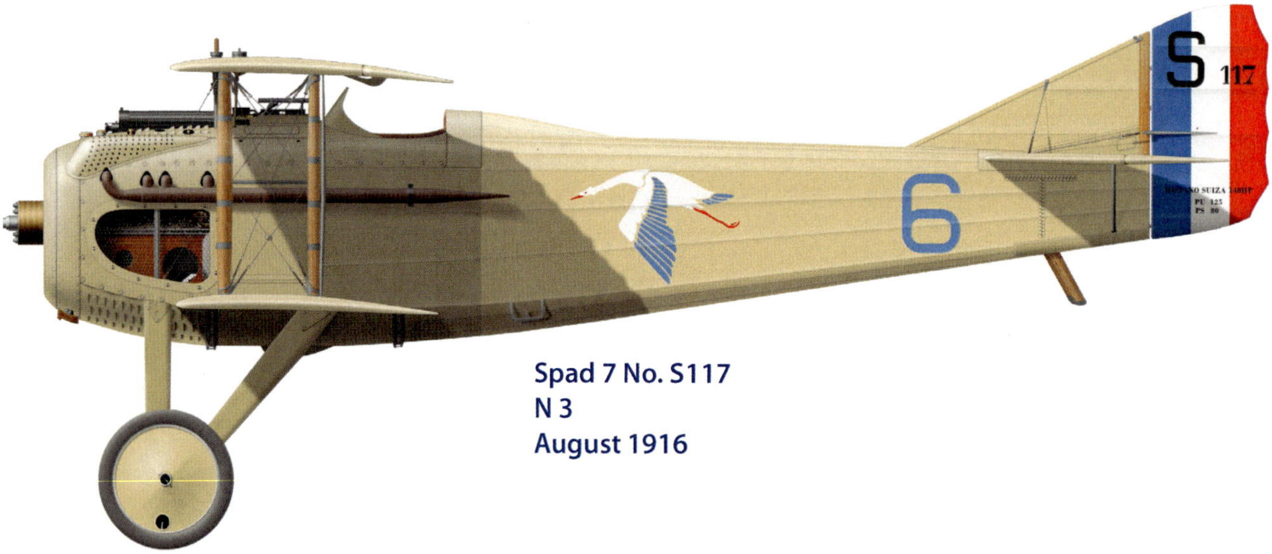

Spad 7 No. S117
N 3
August 1916

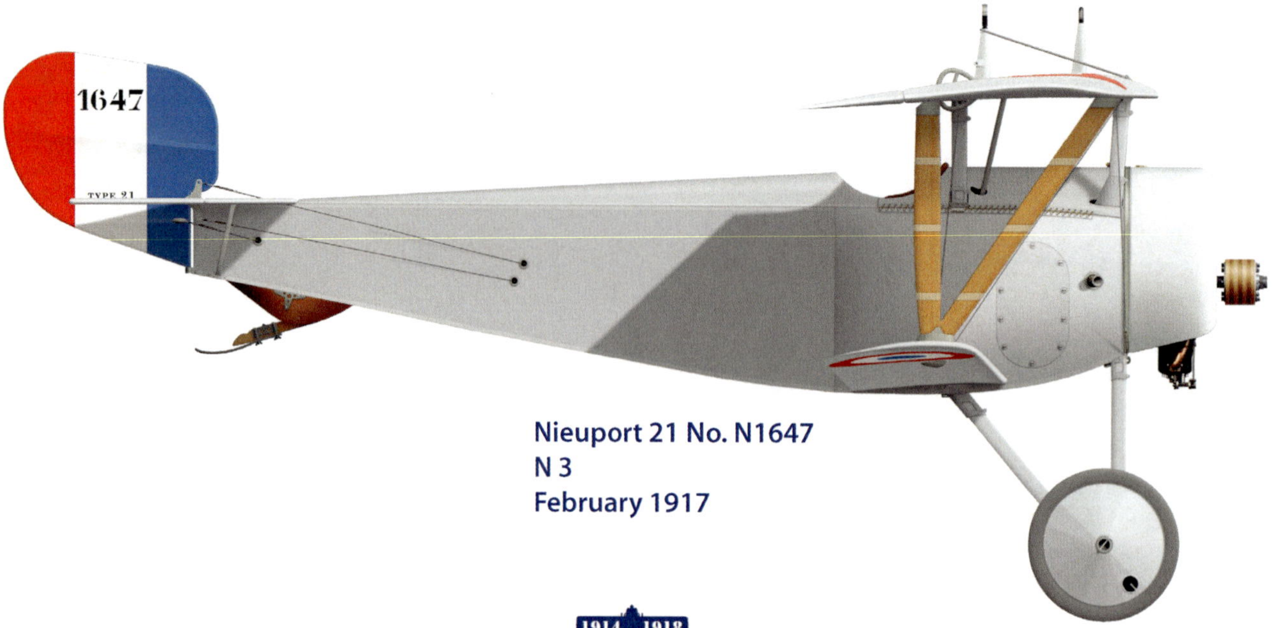

Nieuport 21 No. N1647
N 3
February 1917

Spad 7 *l'Oiseau bleu 6* No. S225
N 3
February 1917

Morane-Saulnier 406 C1 No. 238 (N600)
GC III/6
September 1939

Morane-Saulnier 406 C1 No. 687 (L716)
GC III/6
Spring 1940

Jean Chaput
(1893–1918)
Lieutenant
MS 26, C 28, MF 28, N 57, Spa 57
16 victories

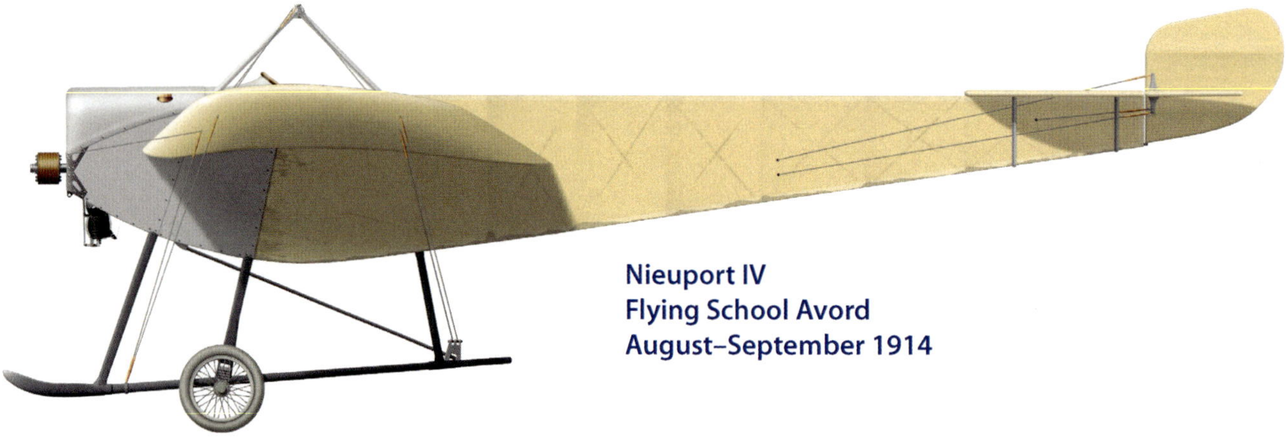

Nieuport IV
Flying School Avord
August–September 1914

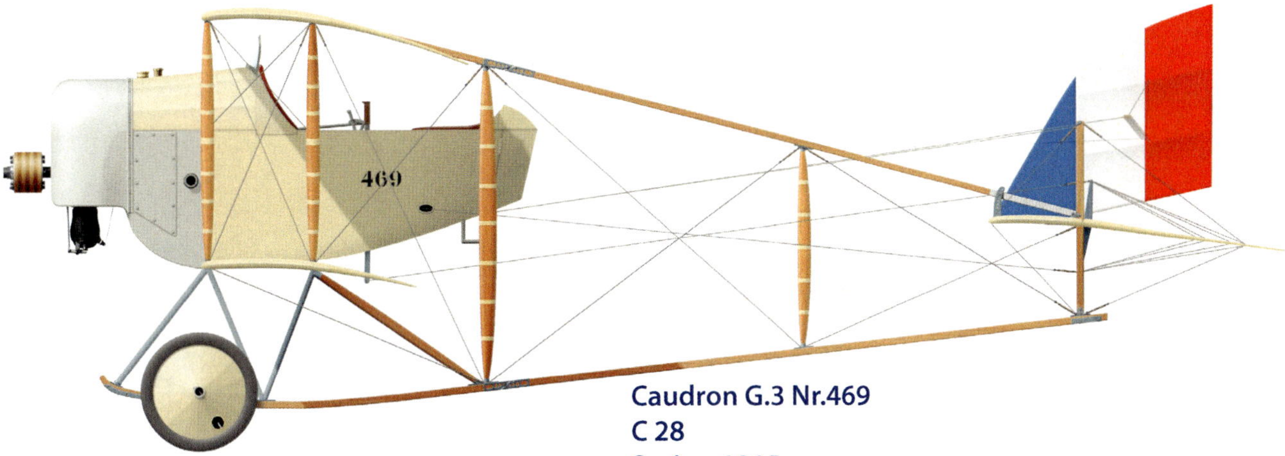

Caudron G.3 Nr.469
C 28
Spring 1915

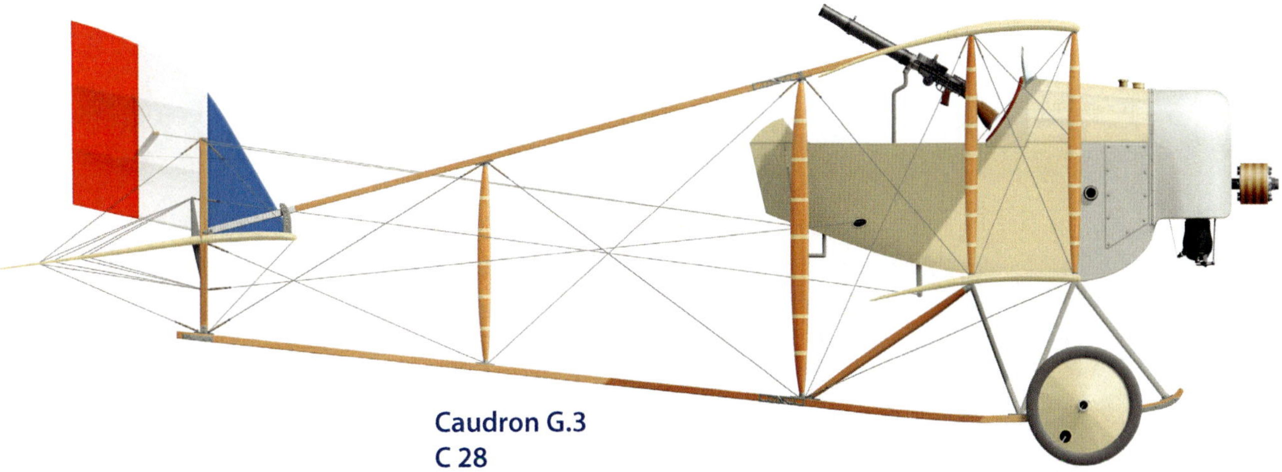

Caudron G.3
C 28
July 1915

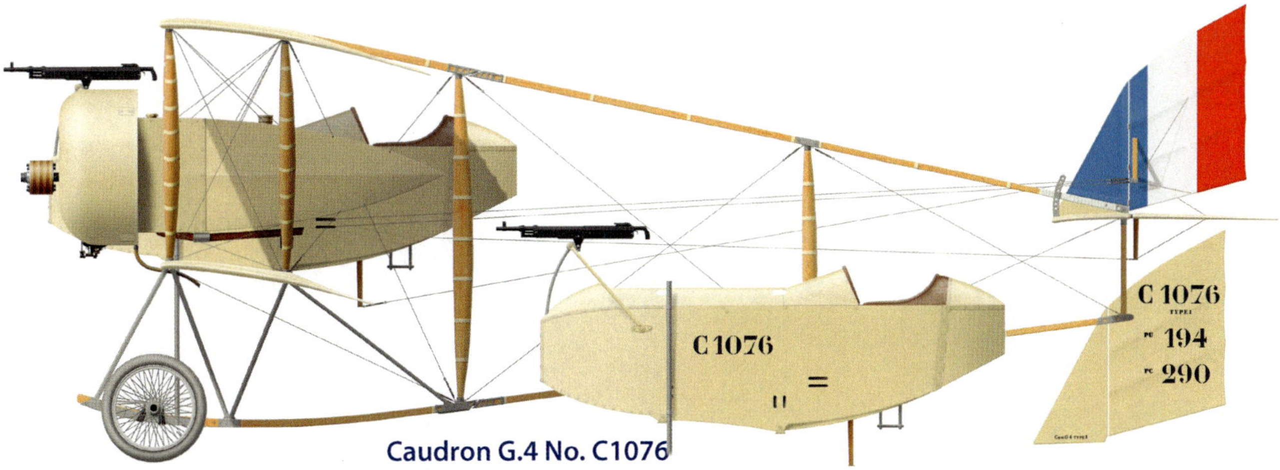

Caudron G.4 No. C1076
C 28
November 1915

Morane-Saulnier L No. MS353
C 28
November 1915

Morane-Saulnier N
Groupement d'Angers
March 1916

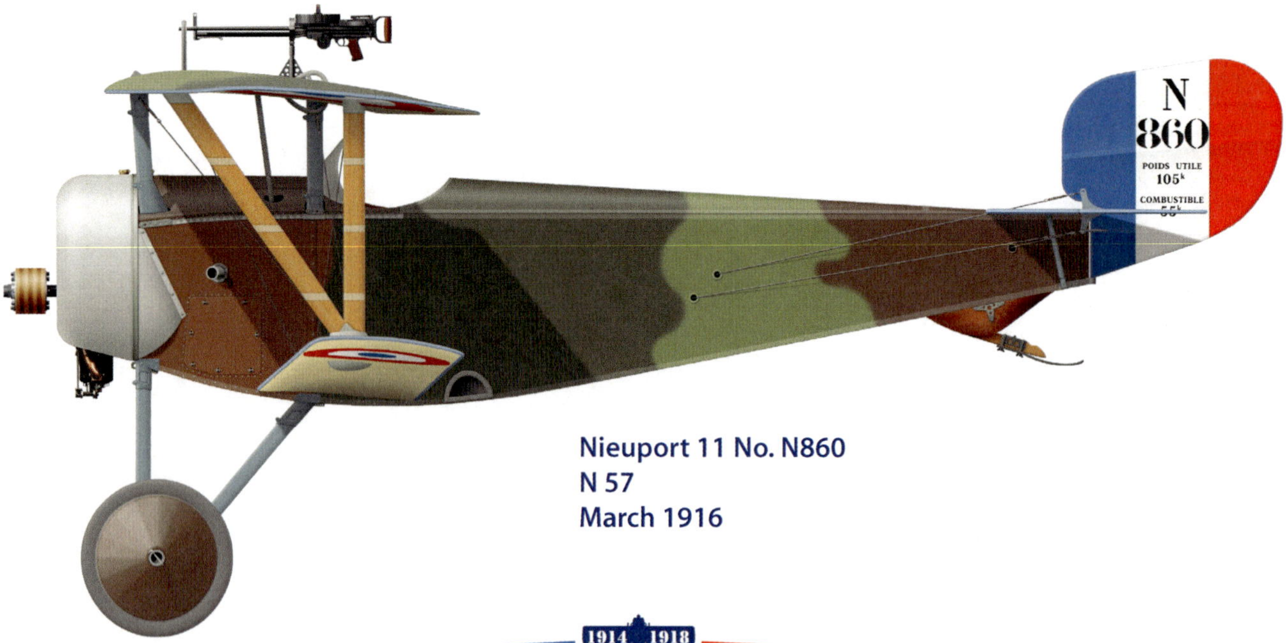

Nieuport 11 No. N860
N 57
March 1916

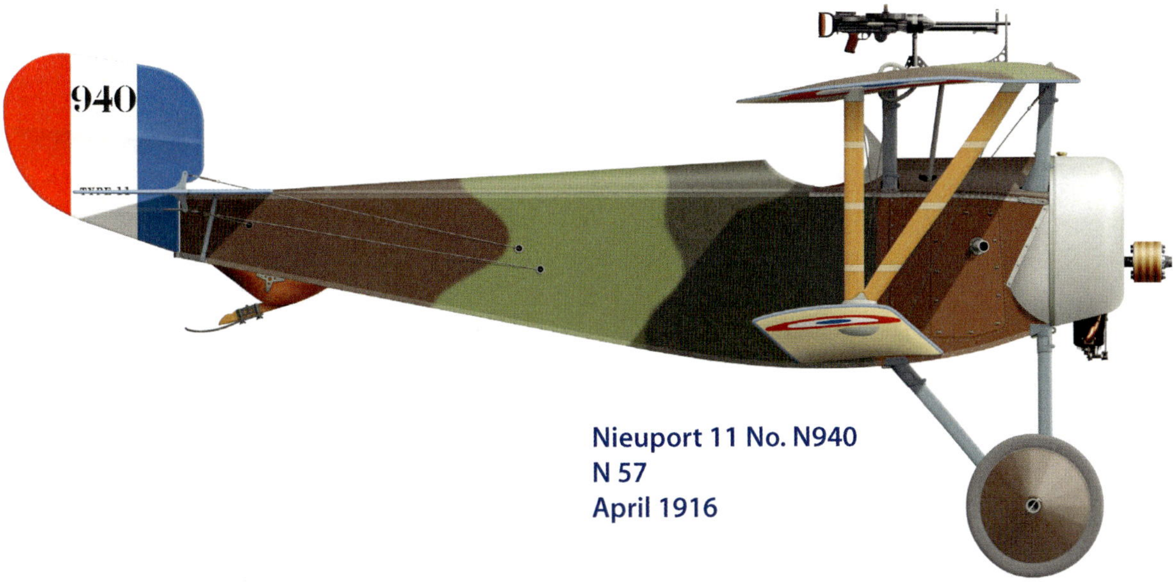

Nieuport 11 No. N940
N 57
April 1916

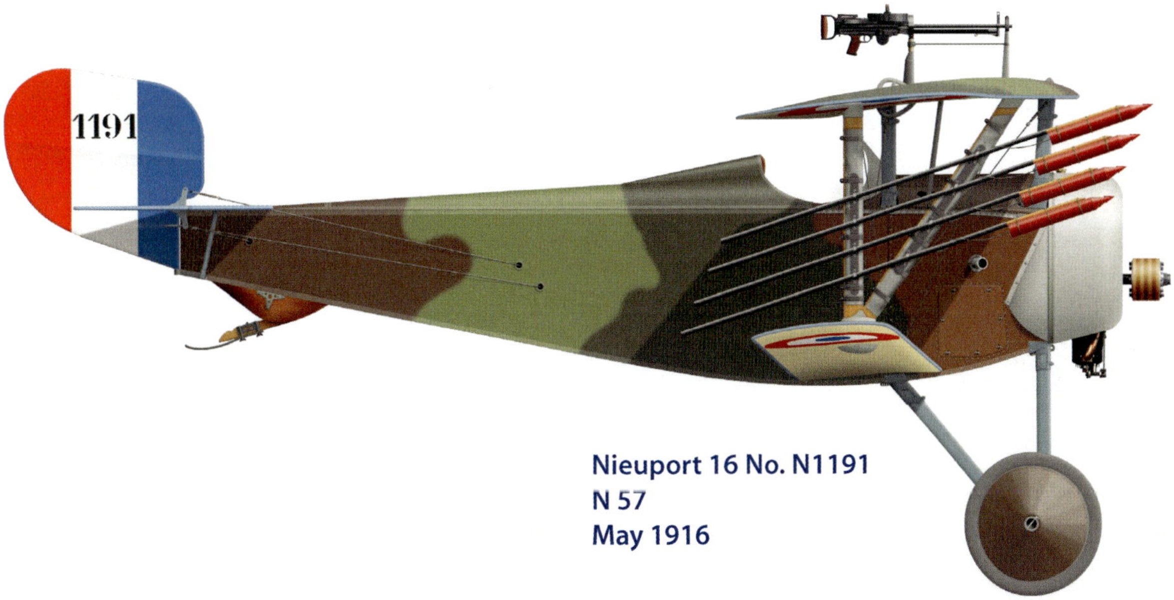

Nieuport 16 No. N1191
N 57
May 1916

Nieuport 17 No. N2280
N 57
April 1917

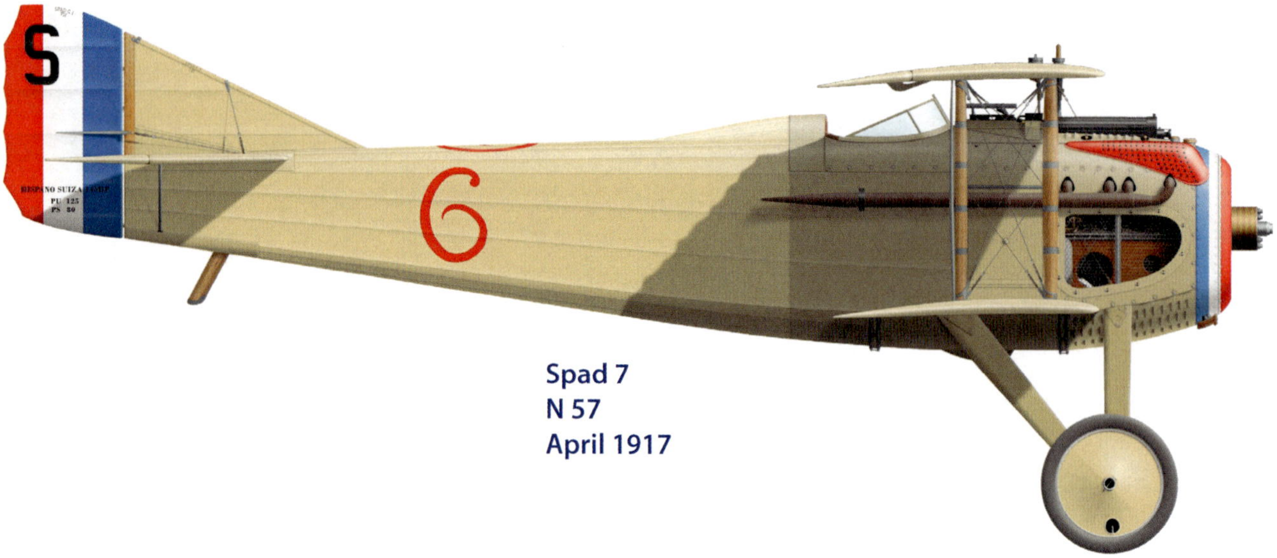

Spad 7
N 57
April 1917

Spad 7
N 57
March 1918

Louis-Antoine Chartoire
(1895–1992)
Sous-Lieutenant
N 31, Spa 31
5 victories

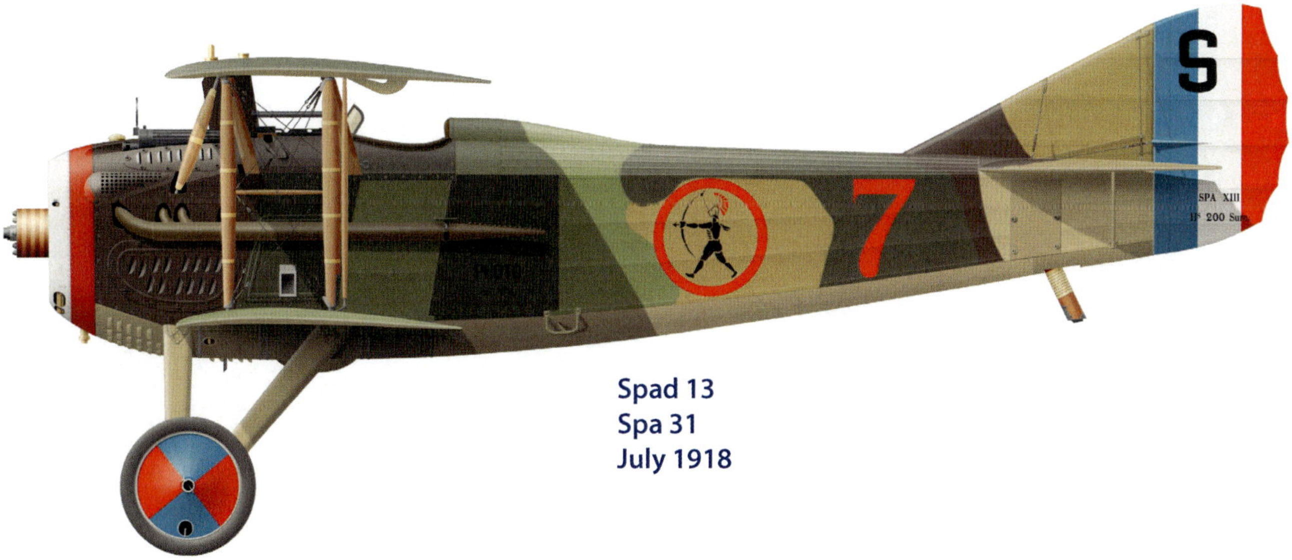

Spad 13
Spa 31
July 1918

Fernand Henri Chavannes
(1897–1985)
Sous-Lieutenant
N 112, Spa 112, Spa 176
7 victories

Voisin LA 5 No. V.1311
Military Flying School Ambrieu
Autumn 1916

Voisin LA 5 No. V.534
Military Flying School Ambrieu
Autumn 1916

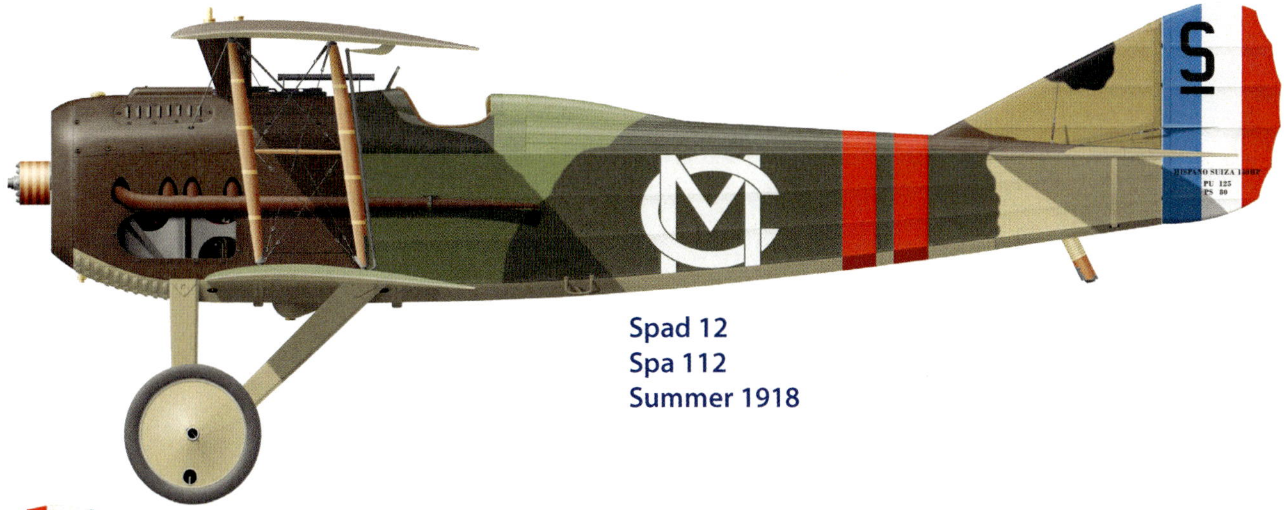

Spad 12
Spa 112
Summer 1918

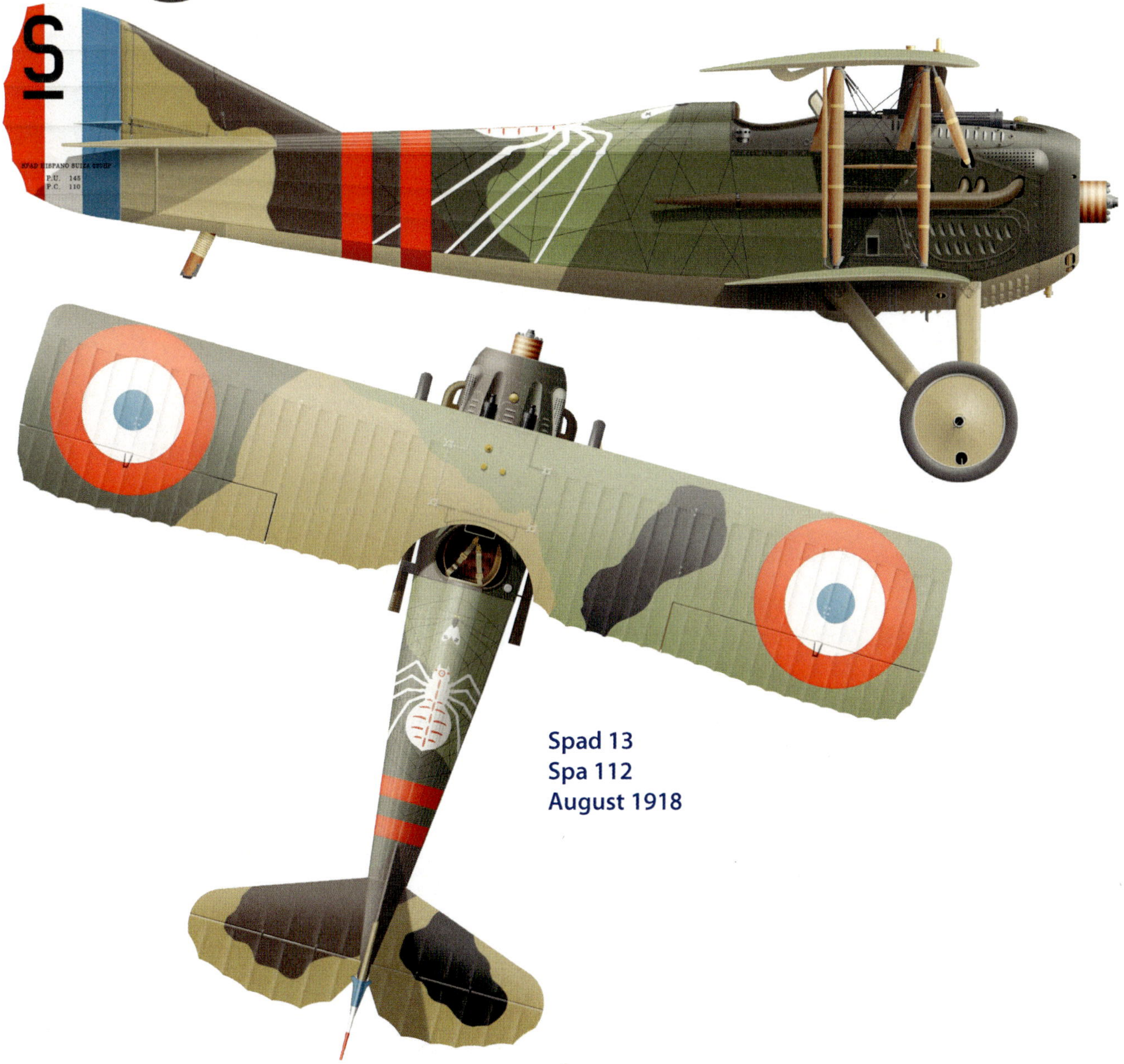

Spad 13
Spa 112
August 1918

Arthur Marie Marcel Coadou

(1894–1985)
Sous-Lieutenant
C 9, N 88, Spa 88
7 or 9 victories

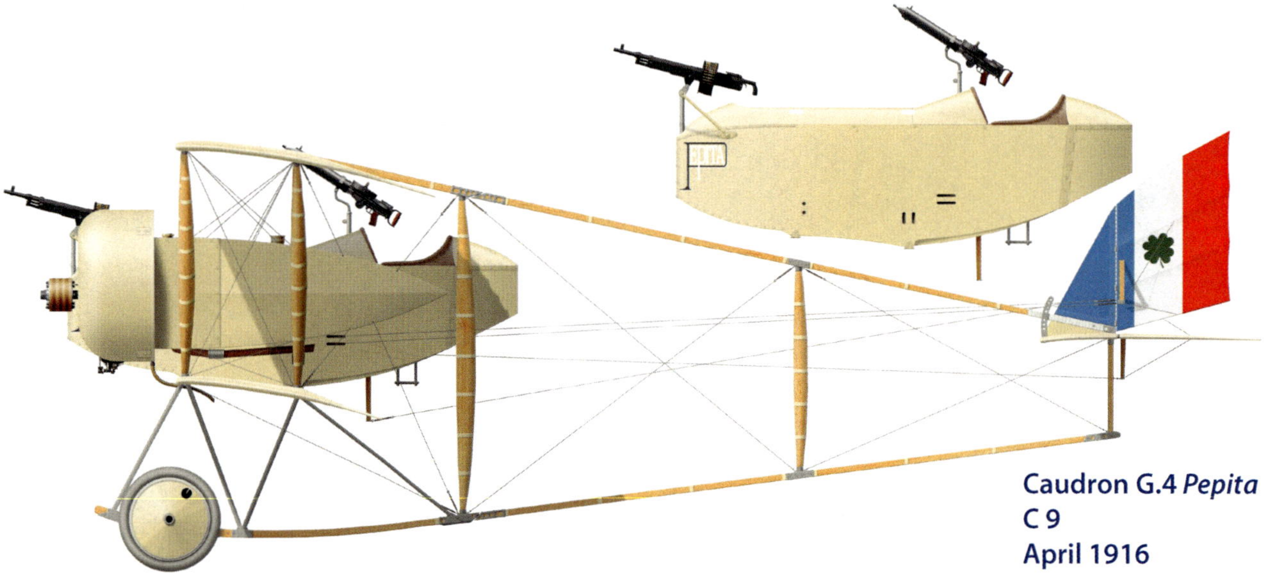

Caudron G.4 *Pepita*
C 9
April 1916

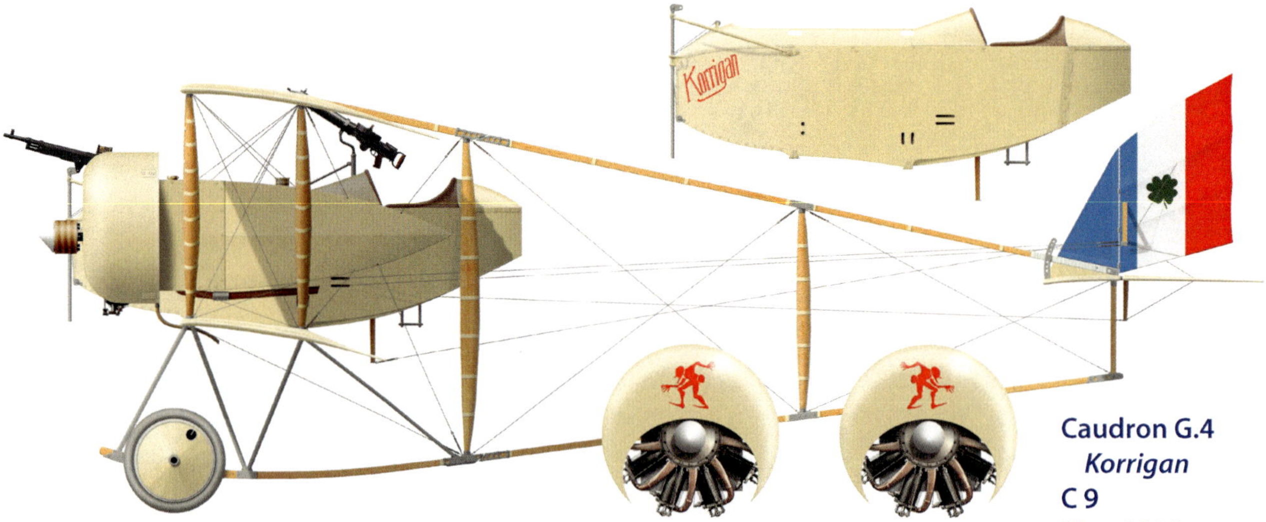

Caudron G.4
Korrigan
C 9
May 1916

Nieuport 24*bis* *Judex* No. N3133
N 88
Summer 1917

Spad 7 *Judex* No. S1369
N 88
Autumn 1917

Spad 13 No. S15654
Spa 88
November 1918

Morane-Saulnier MS 406C1 '7' No.438
2e Escadrille GC I/2
April 1940

Michel Joseph Callixte Marie Coiffard
(1892–1918)
Lieutenant
N 154, Spa 154
34 victories

Nieuport 24*bis* No. N4473
N 154
1917

Nieuport 24*bis* No. N4451
N 154
1917

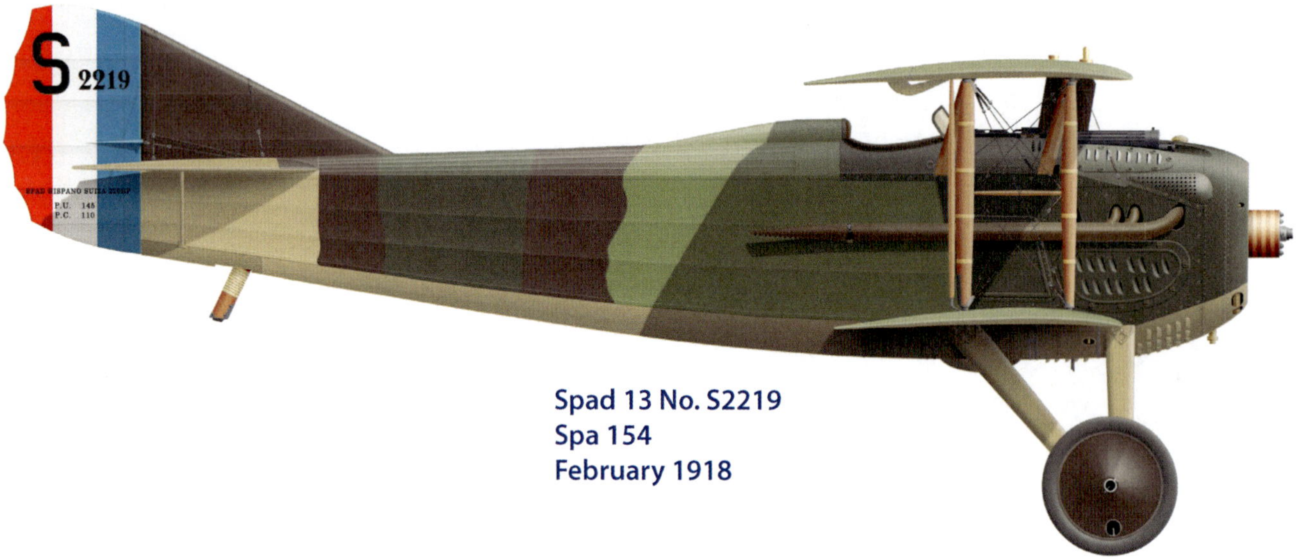

Spad 13 No. S2219
Spa 154
February 1918

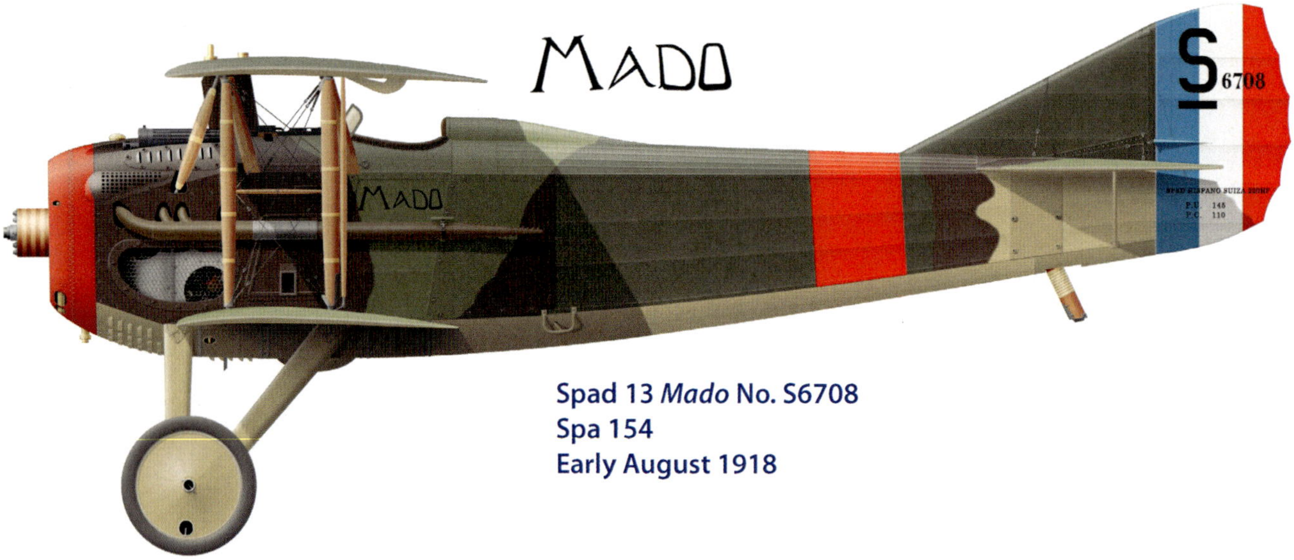

Spad 13 *Mado* No. S6708
Spa 154
Early August 1918

Spad 13 *Valentine* No. S4830
Spa 154
October 1918

Theophile Henri Condemine
(1895–1975)
Sous-Lieutenant
Spa 154
9 victories

Spad 13
Spa 69
May 1918
Condemine flew this aircraft during a front-line internship before his appointment to Spa 154.

James Connely
(1884–1944)
Sergent
Spa 157, Spa 163
8 victories

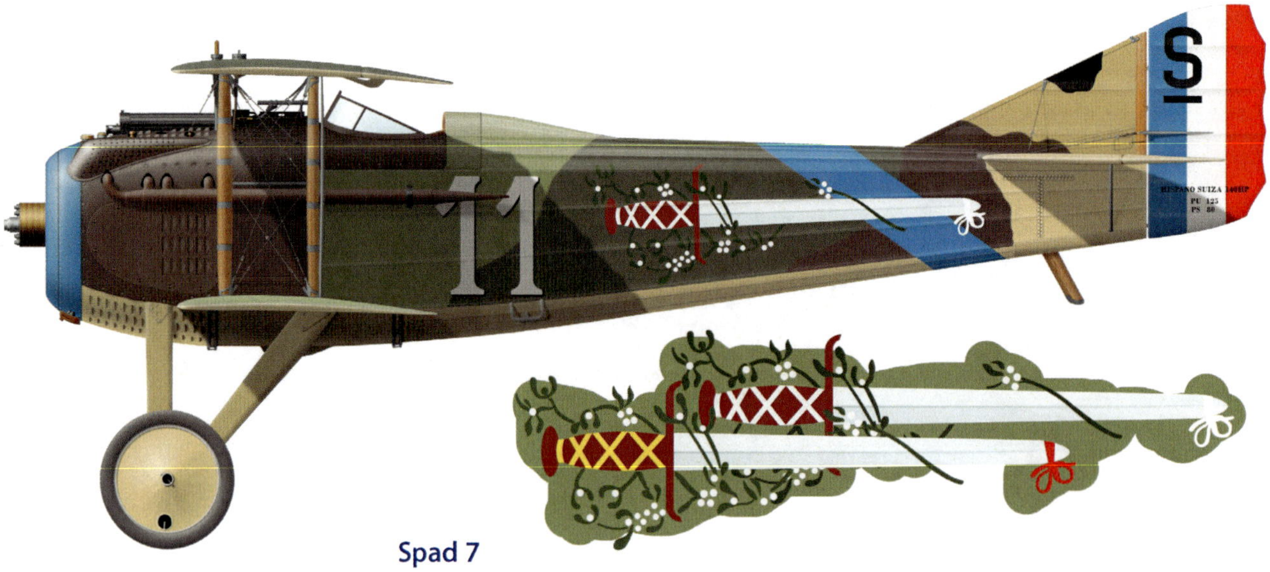

Spad 7
Spa 163
August–September 1918

Diedonné Costes
(1892–1973)
Sous-Lieutenant
MF 55, MF 85 (502), MF 392, N 506, N 507, N 531
8 victories

Maurice Farman 11bis No. MF1076
MF 502
Summer 1916

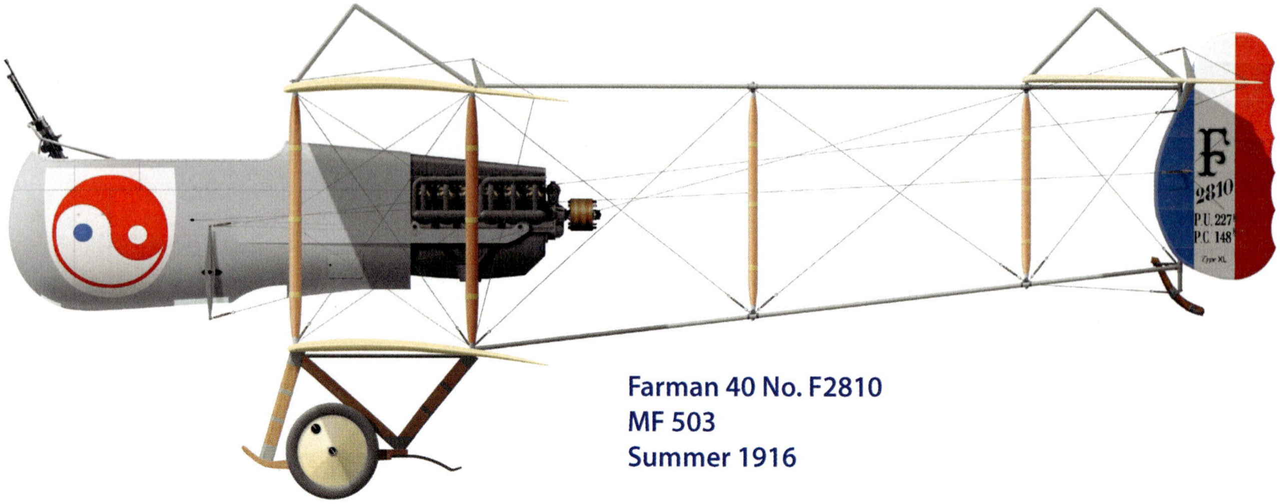

Farman 40 No. F2810
MF 503
Summer 1916

Sopwith 1A2 No.31
N 507
Winter 1916–1917

Nieuport 23 No. N3858
N 507
April 1917

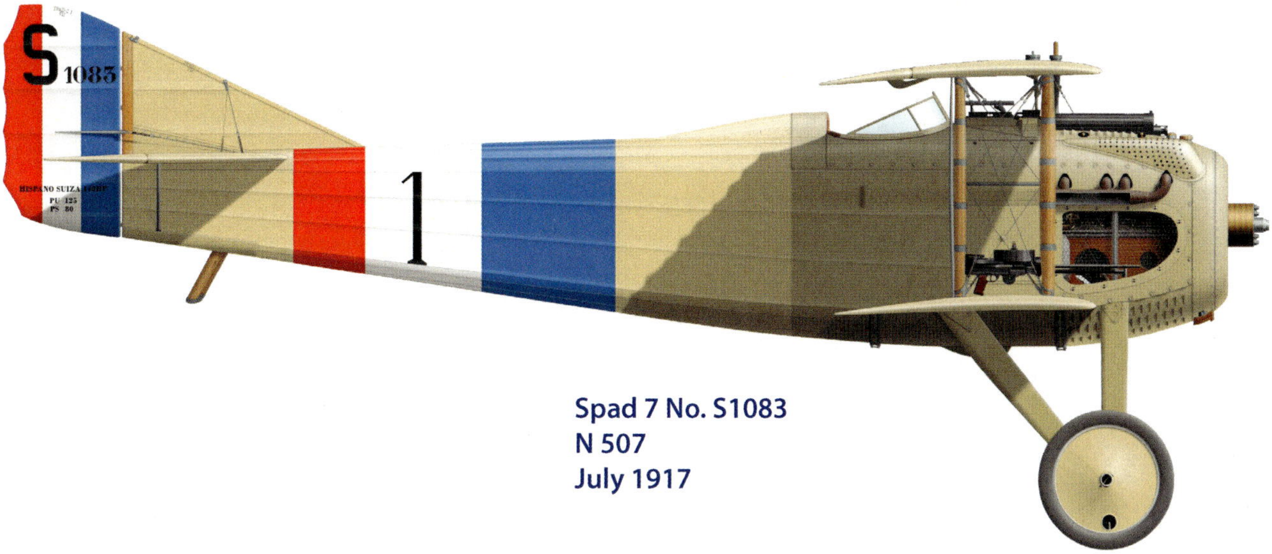

Spad 7 No. S1083
N 507
July 1917

Spad 7 No. S1083
N 507
August 1917

Louis Fernand Coudouret

(1896–1929)
Lieutenant
N 57, N 102, N 112, Spa 103, VB 102, N 581
6 victories

Spad 7
N 581
Summer 1917

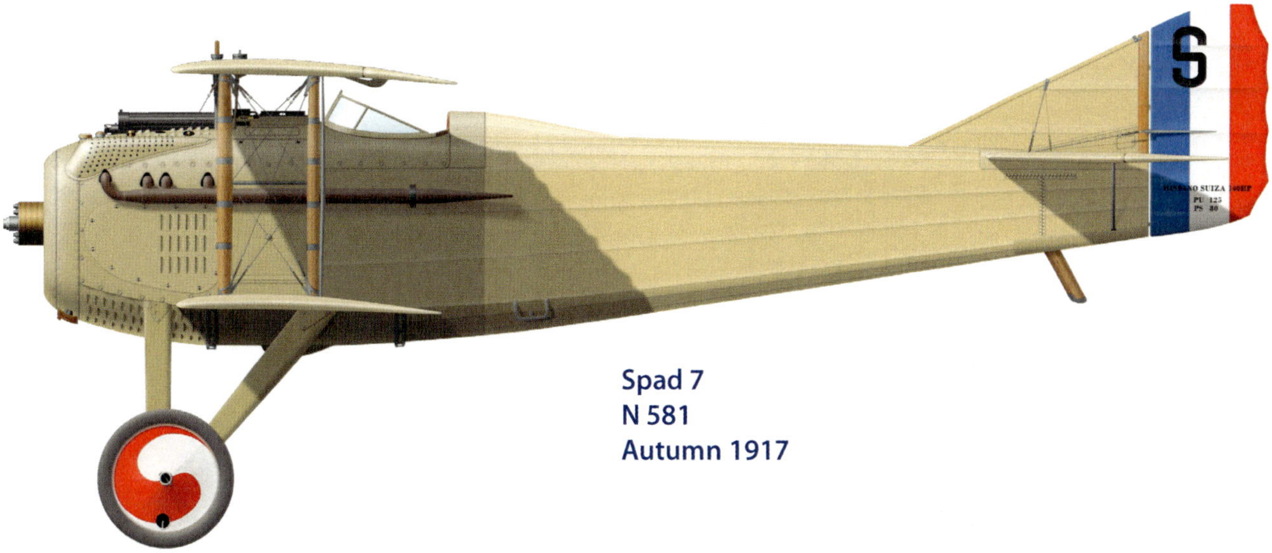

Spad 7
N 581
Autumn 1917

Spad 13 No. S4703
Spa 103
Summer 1918

Gustave Victorine Daladier

(1888–1974)
Lieutenant
F 14, N 73, N 93, Spa 93
12 victories

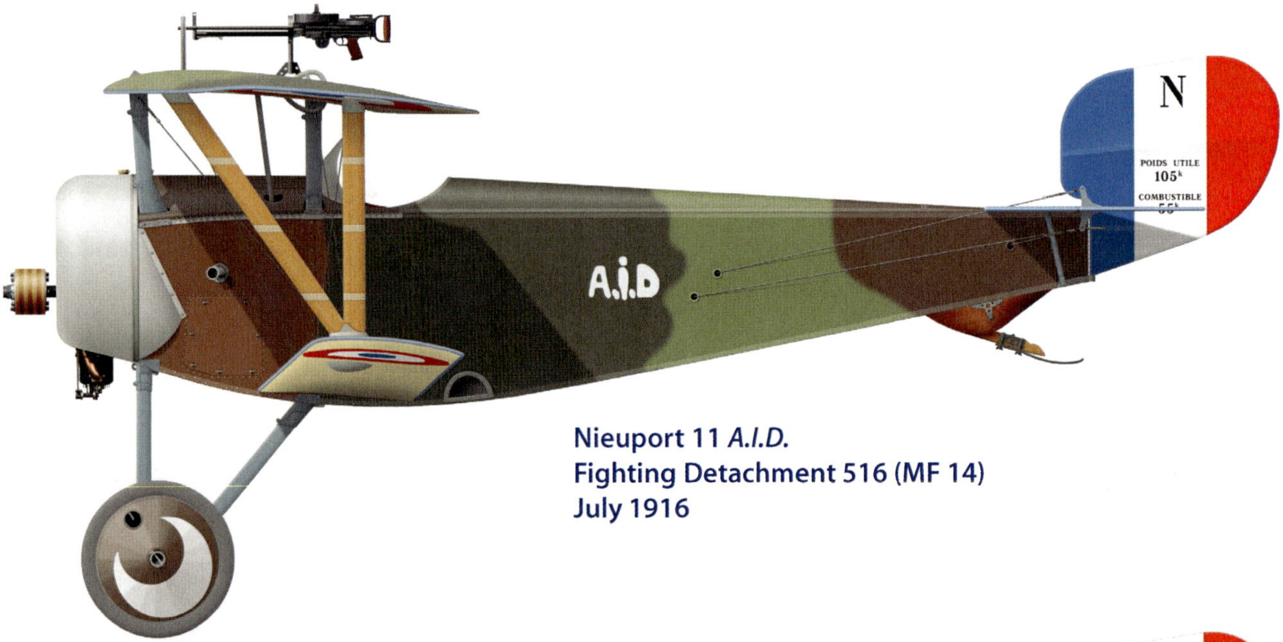

Nieuport 11 *A.I.D.*
Fighting Detachment 516 (MF 14)
July 1916

Nieuport 17
N 93
Summer 1917

Robert Joseph Delannoy
(1890–1979)
Sous-Lieutenant
N 80, Spa 80
7 victories

Spad 13 No. S2287
Spa 80
Spring 1918

Omer Paul Demeuldre
(1892–1918)
Sous-Lieutenant
MF 63, Spa 84
13 victories

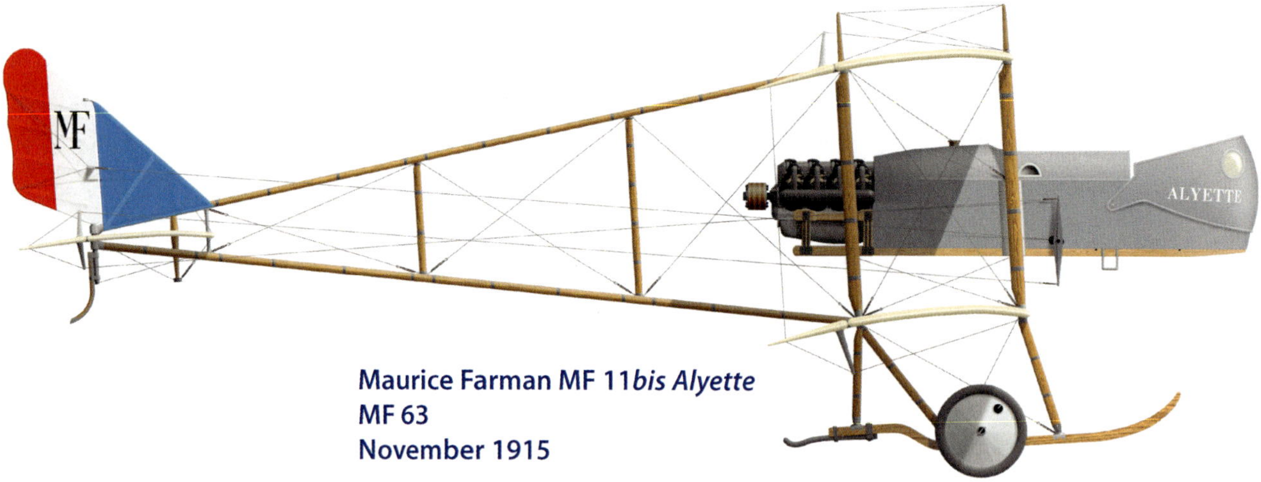

Maurice Farman MF 11*bis Alyette*
MF 63
November 1915

Sopwith 1A2 *III KISS-ME* NO. N4048
MF 63
Summer 1917

Spad 7 *KISS-ME*
Spa 84
Spring 1918

Spad 7
Spa 84
Spring 1918

Jean Marie Emile Derode
(1887–1918)
Capitaine
N 67, N 102, Spa 99
7 victories

Nieuport 16 No.1130
N 67
17 June 1916
This was Navarre's plane. Navarre was wounded 17 June 1916 and Derode, flying this plane that day, crashed it on landing.

Spad 7 No. S1102
Spa 102
Summer 1917

Spad 7
Spa 102
Summer 1917

Spad 13
Spa 99
February–March 1918
Reconstruction

Albert Louis Deullin
(1890–1923)
Capitaine
MF 62, N 3, Spa 73
20 victories

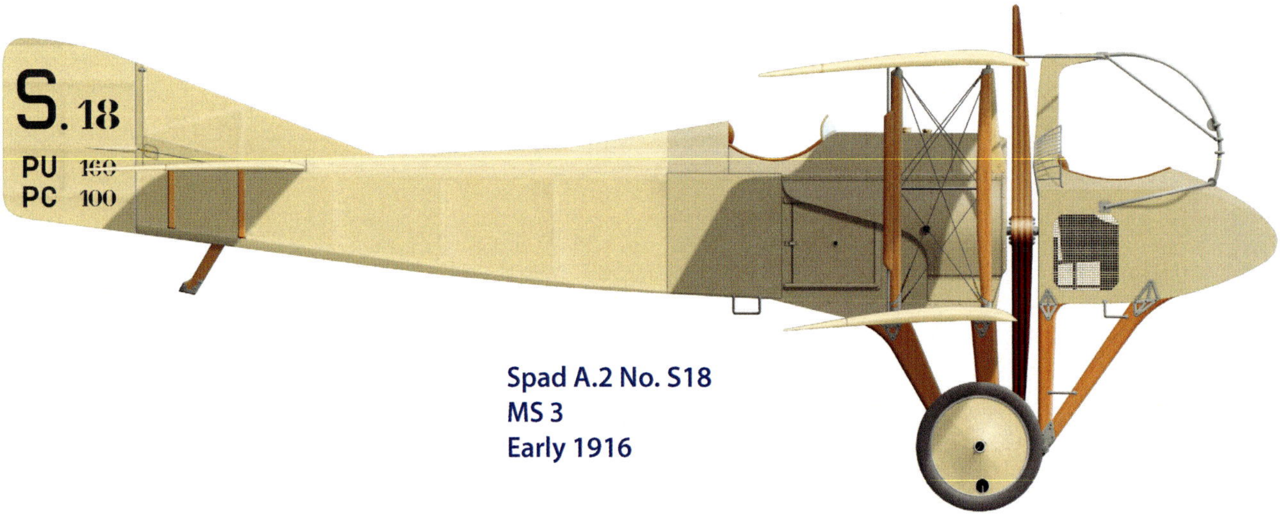

Spad A.2 No. S18
MS 3
Early 1916

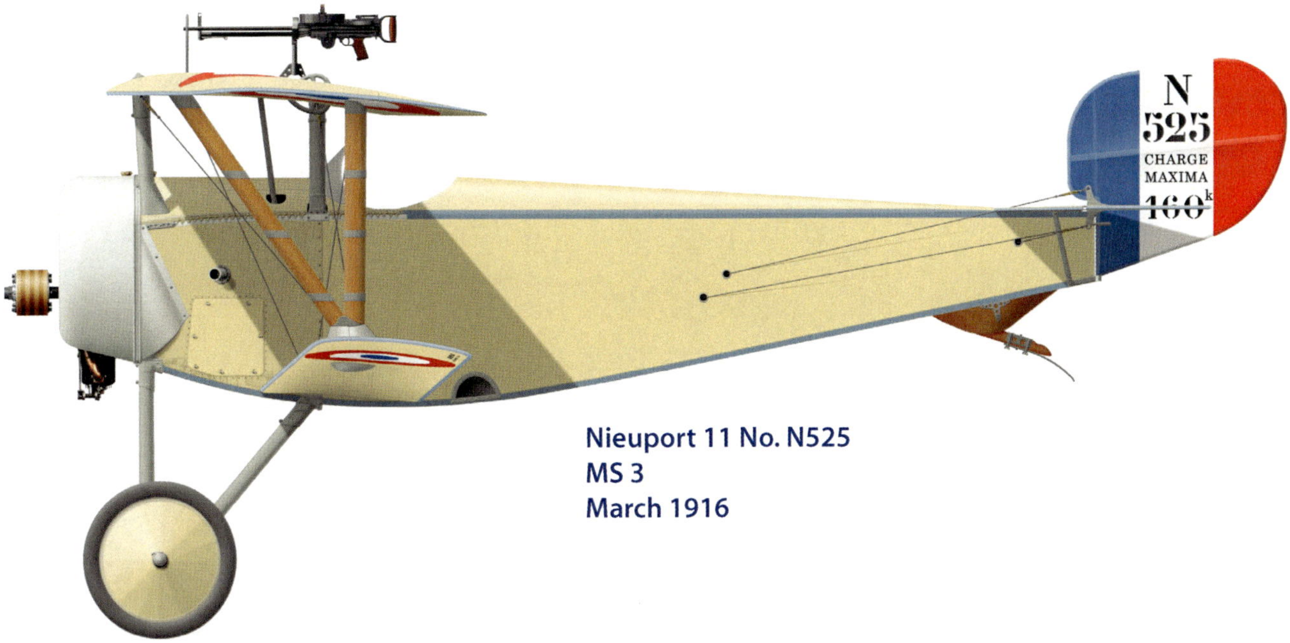

Nieuport 11 No. N525
MS 3
March 1916

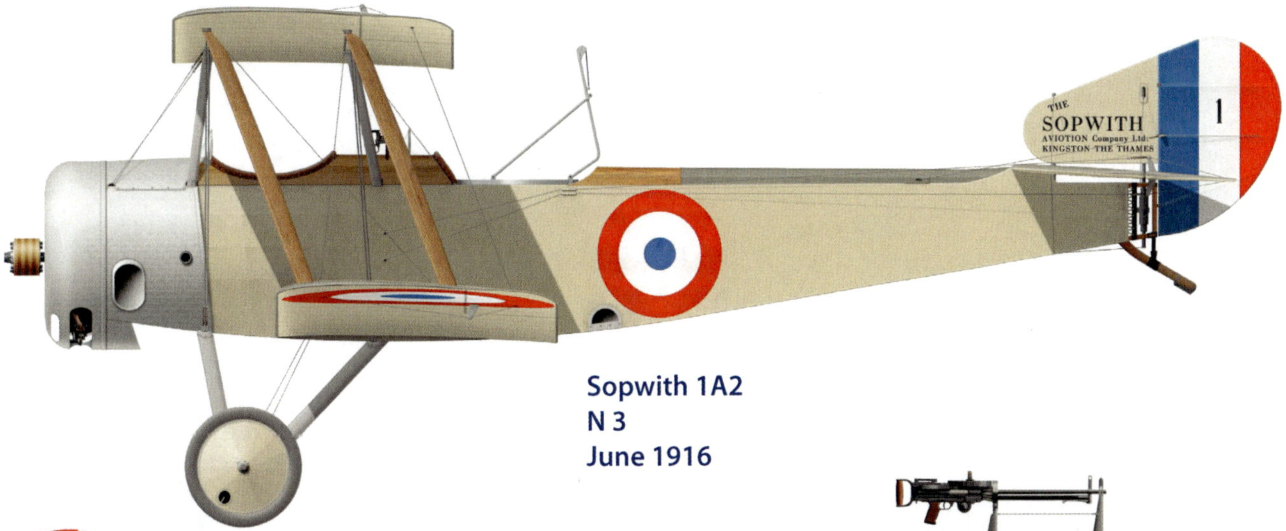

Sopwith 1A2
N 3
June 1916

Nieuport 16 No. N962
N 3
July 1916

Nieuport 17 No. N1532
N 3
Autumn 1916

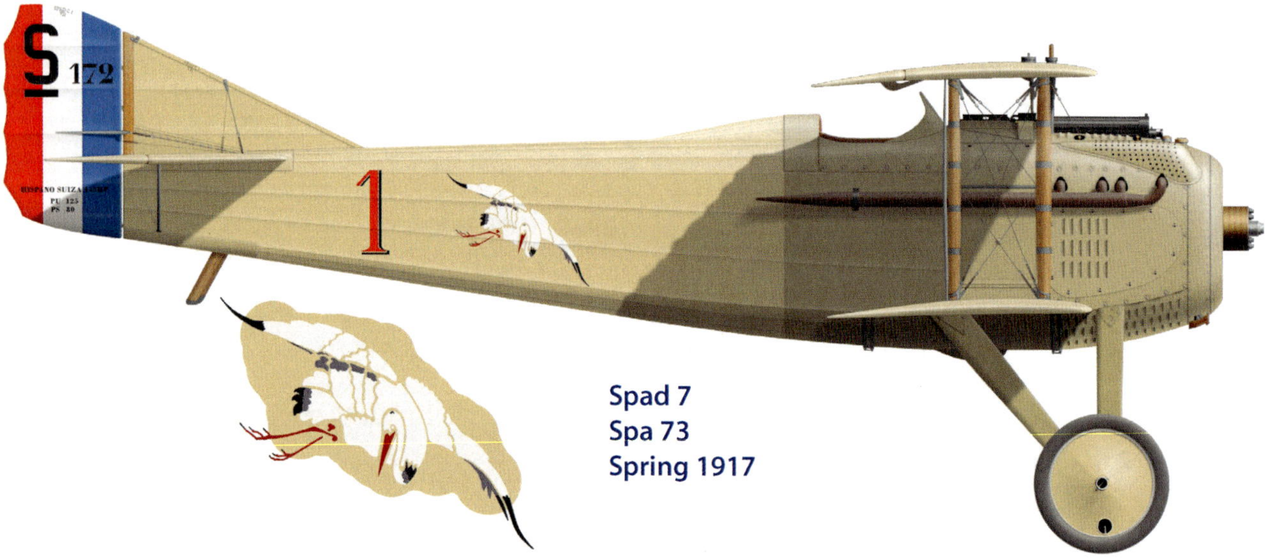

Spad 7
Spa 73
Spring 1917

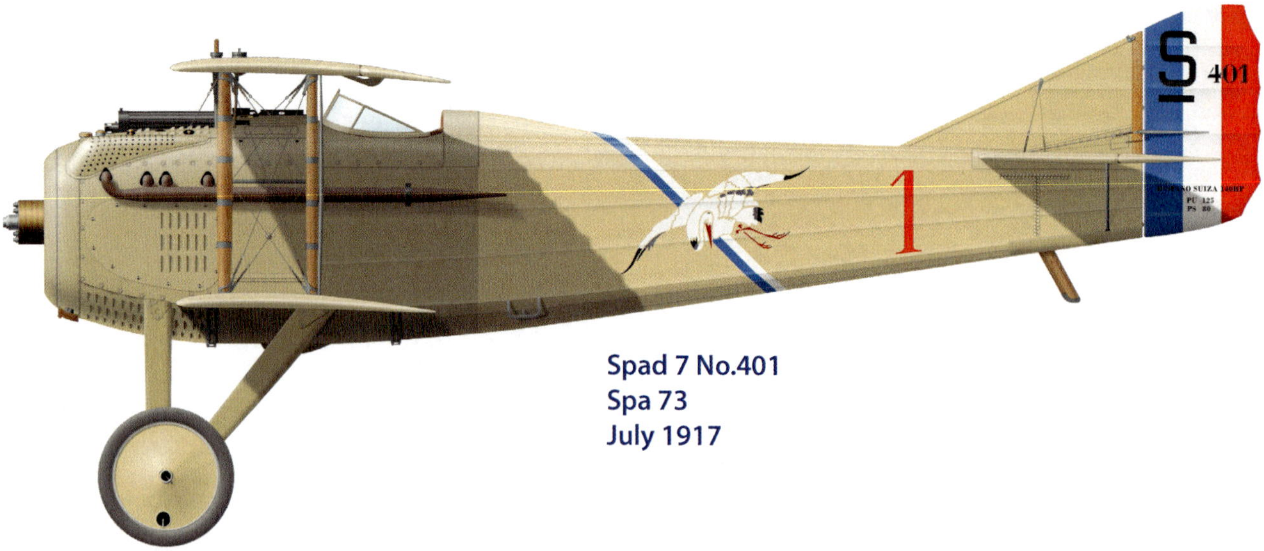

Spad 7 No.401
Spa 73
July 1917

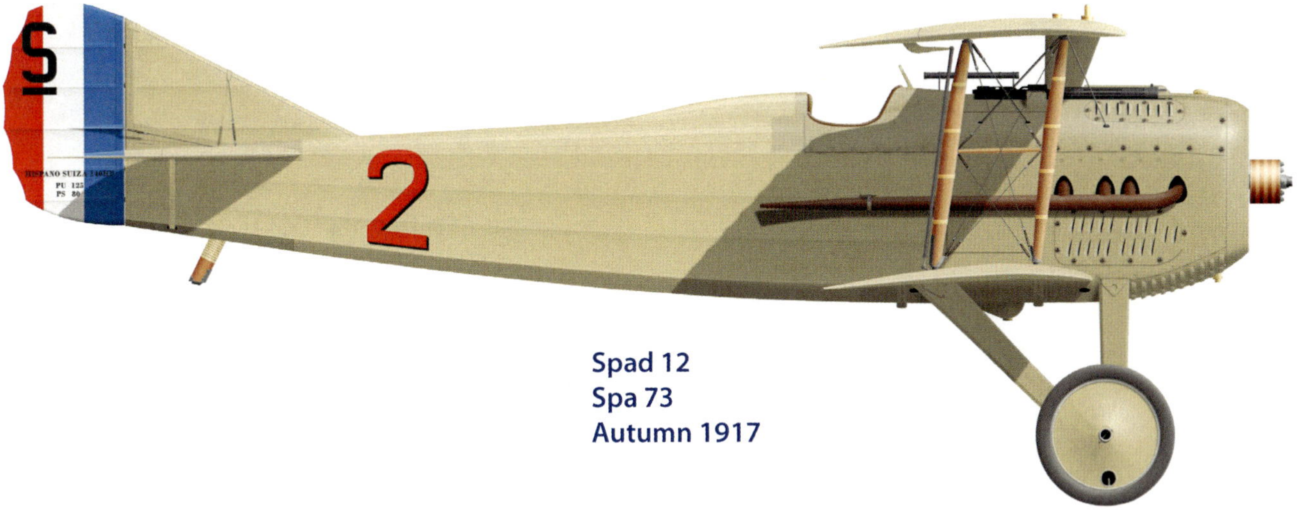

Spad 12
Spa 73
Autumn 1917

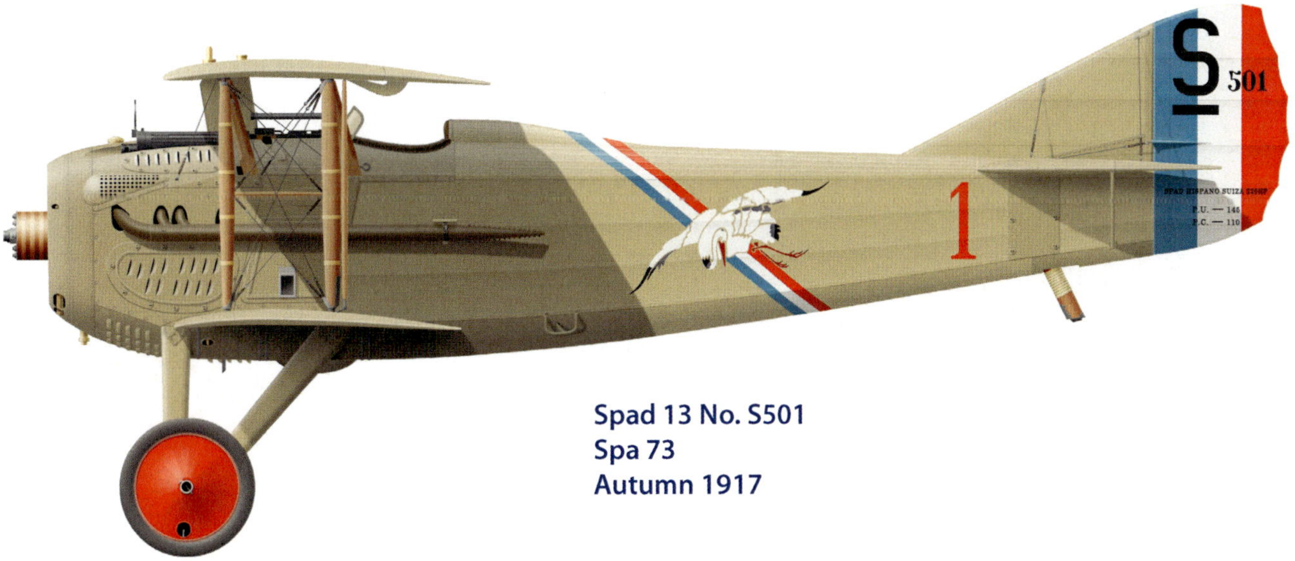

Spad 13 No. S501
Spa 73
Autumn 1917

Marcel Marc Dhôme
(1893–1960)
Sous-Lieutenant
N 81, Spa 81
9 victories

Spad 7 *LULU II*
Spa 81
Late 1917

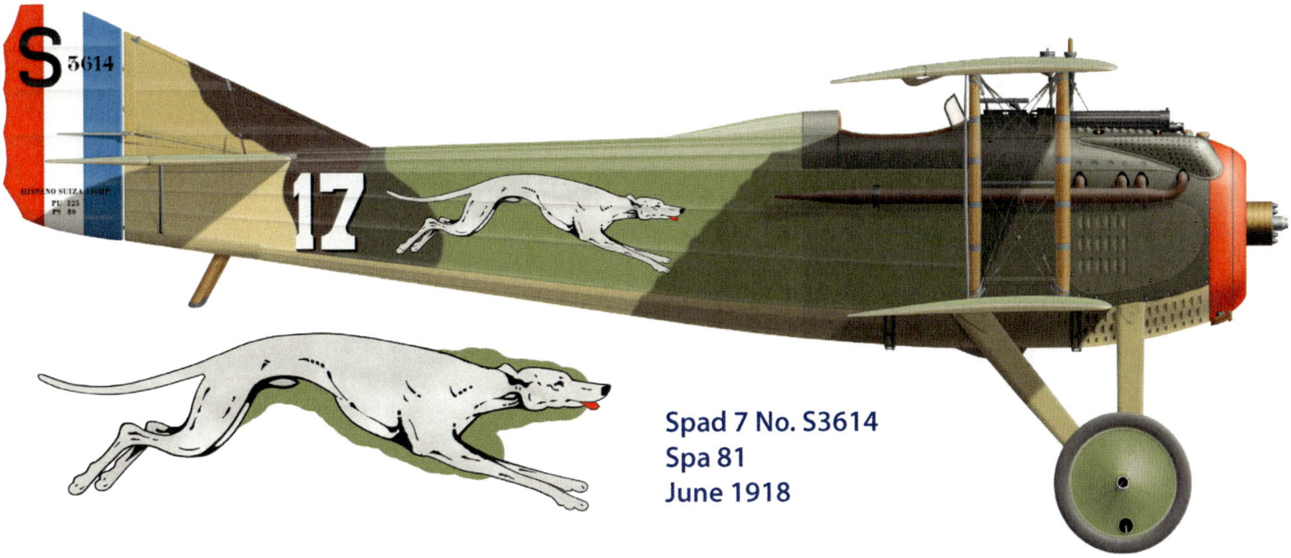

Spad 7 No. S3614
Spa 81
June 1918

Spad 13
Spa 81
Autumn 1918

René Pierre Marie Dorme
(1894–1917)
Sous-Lieutenant
C 94 CRP, N 3
23 victories

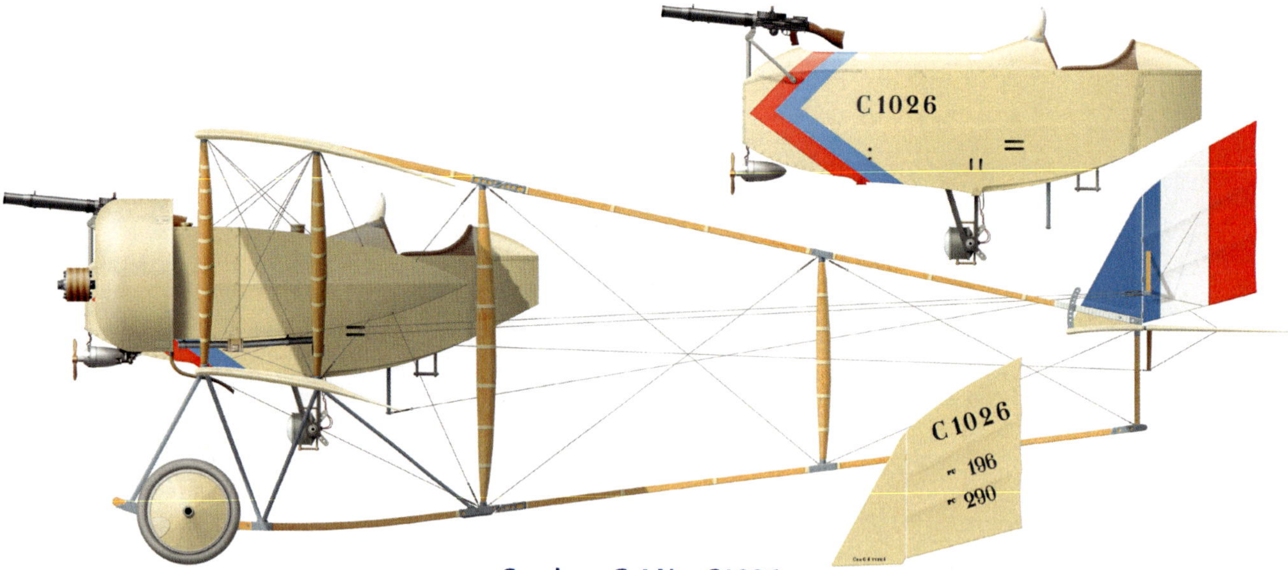

Caudron G.4 No. C1026
C 94 CPR
November 1915

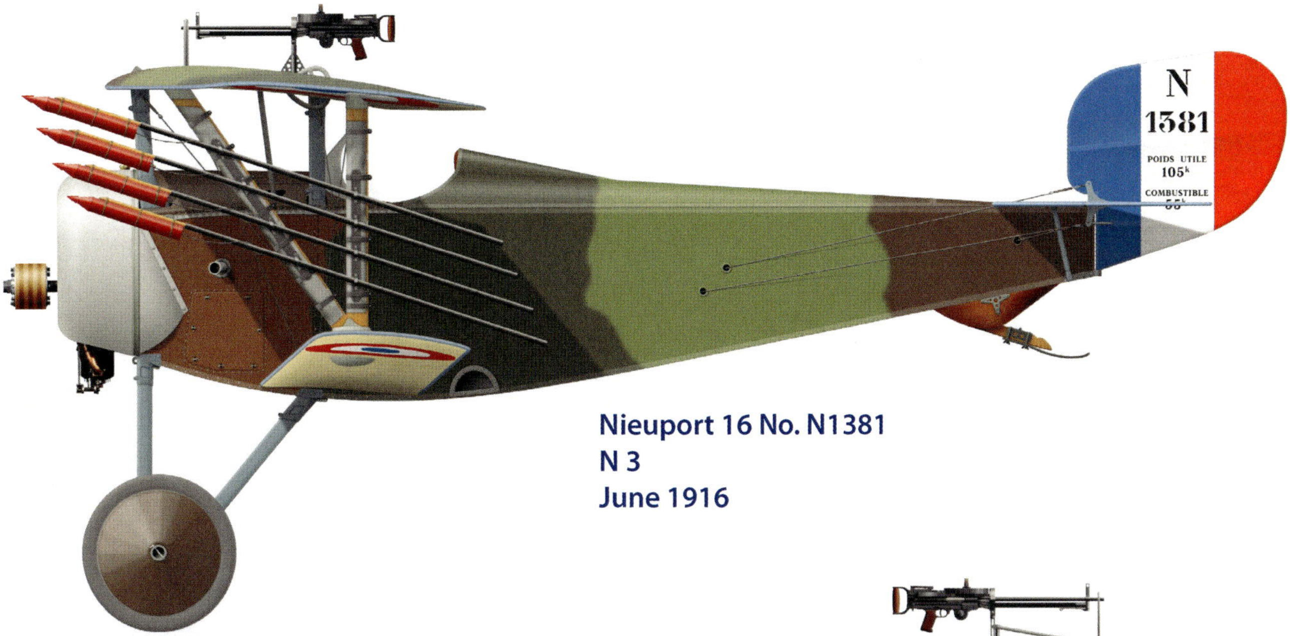

Nieuport 16 No. N1381
N 3
June 1916

Nieuport 16 No. N939
N 3
Summer 1916

Nieuport 17 *Père Dorme II* No. N1428
N 3
July–September 1916

Nieuport 17 *Père Dorme II* No. N1428
N 3
July–September 1916

Nieuport 17 *Père Dorme 3* No. N1720
N 3
December 1916

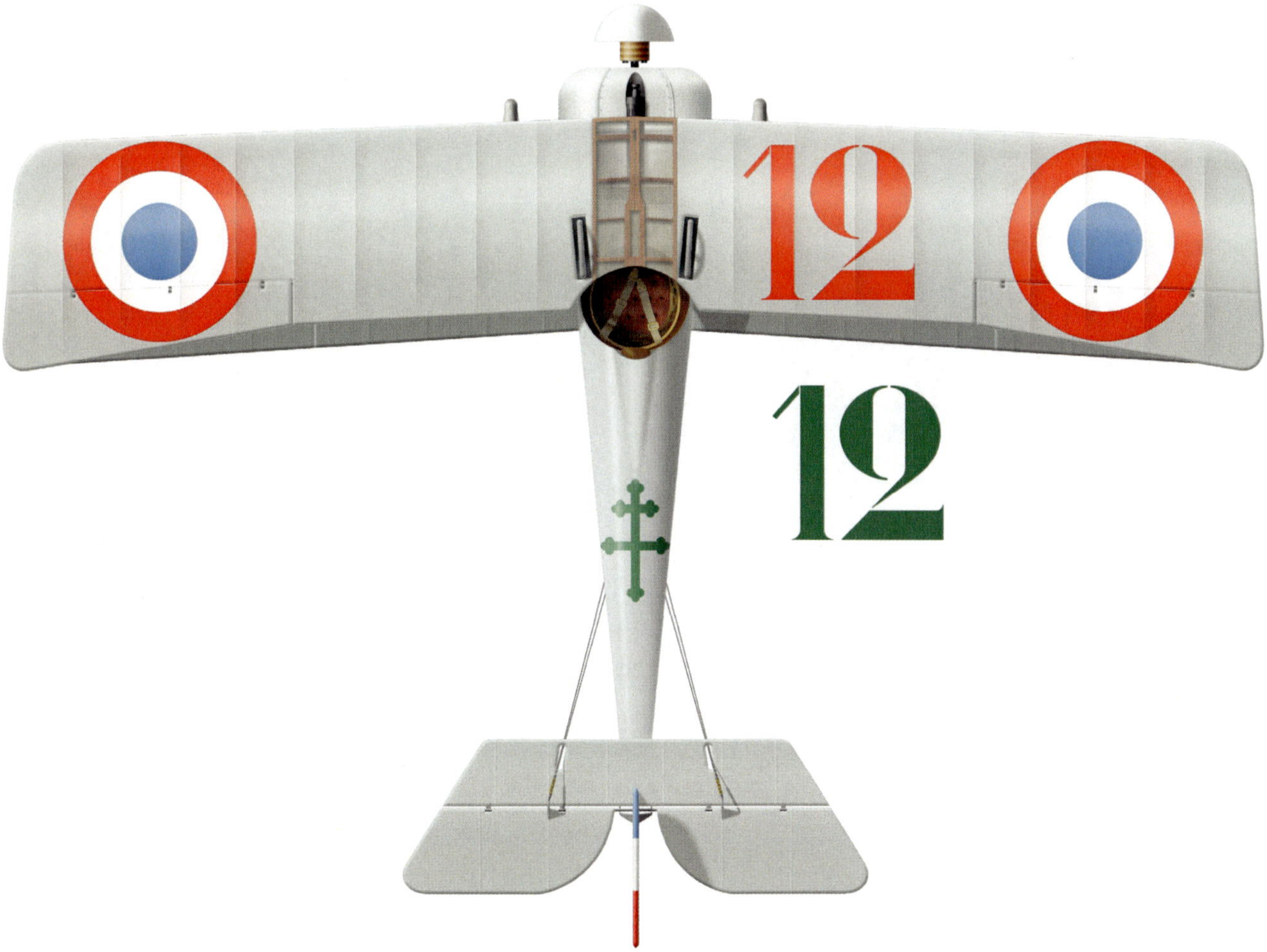

Spad 7 *Père Dorme IV* No. S314
Spa 3
February–March 1917

Spad 7 *Père Dorme* No. S392
Spa 3
Spring 1917

Gustav Augustin Douchy

(1893–1943)
Adjudant
N 38, Spa 38
9 victories

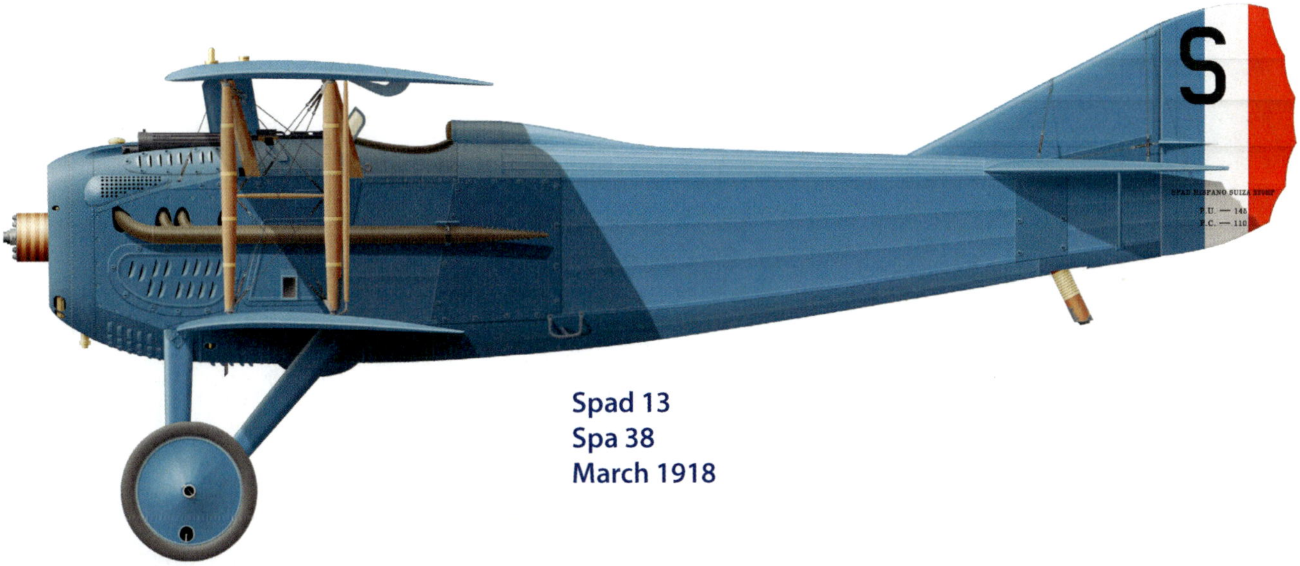

Spad 13
Spa 38
March 1918

Spad 13 No. S2078
Spa 38
March 1918
This plane was also flown
by adjudant Georges Babo.

René Paul Louis Dousinelle
(1891–1963)
Sous-Lieutenant
N311, Spa48
9 victories

Spad 7
Spa 48
October 1917

Spad 13
Spa 48
Summer 1918

André Dubonnet
(1897–1980)
Marechal-des-Logis
Spa 3
6 victories

Spad 13 No. S8174
Spa 3
December 1917

Morane-Saulnier MS 406C1 No. 678 (L-707)
Escadrille 1, GC I/2
October 1940

Pierre Desire Augustin Ducornet
(1898–1963)
Marechal-des-Logis
N 93, Spa 93
7 victories

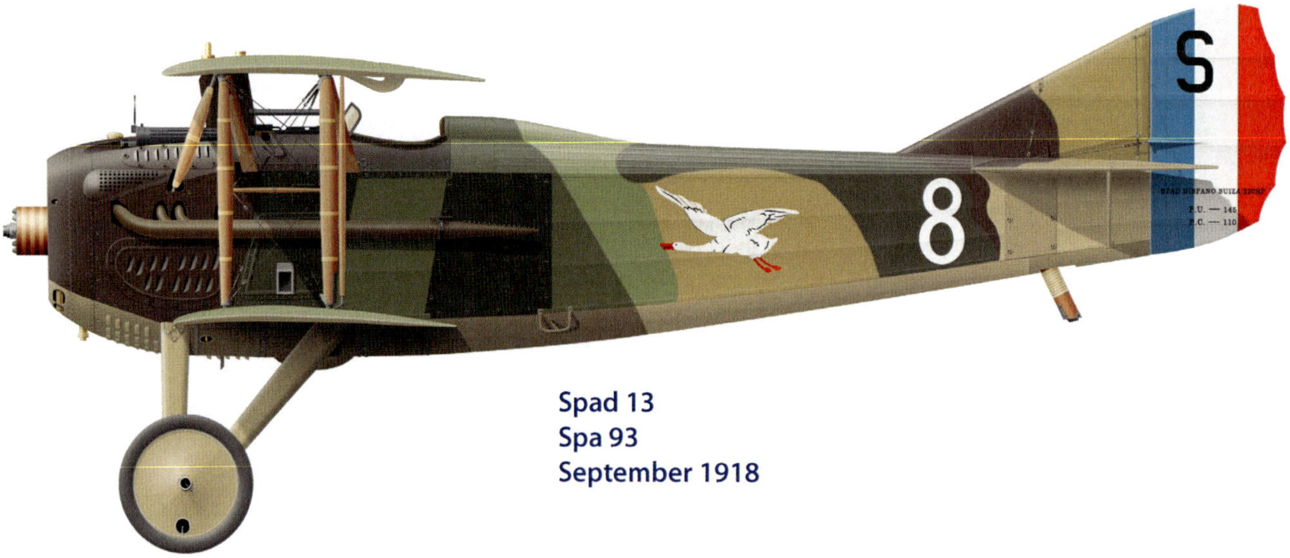

Spad 13
Spa 93
September 1918

Pierre Henri Edmond Dufaur de Gavardie (1890–1966)

Lieutenant
C 53, N 12
6 victories

Nieuport 11 No. N642
N 12
Summer 1916

Nieuport 17 No. N184?
N 12
October 1916

Nieuport 17 No. N181?
N 38
October 1916

Raoul Cesar Robert Pierre Echard

(1883–1922)
Captaine
N12, N26, N82
7 victories

Spad 7 No. S257
N 82
May–July 1917

Nieuport 24
N 82
February 1918

Jacques Louis Ehrlich
(1893–1953)
Sous-Lieutenant
Spa 154
19 victories

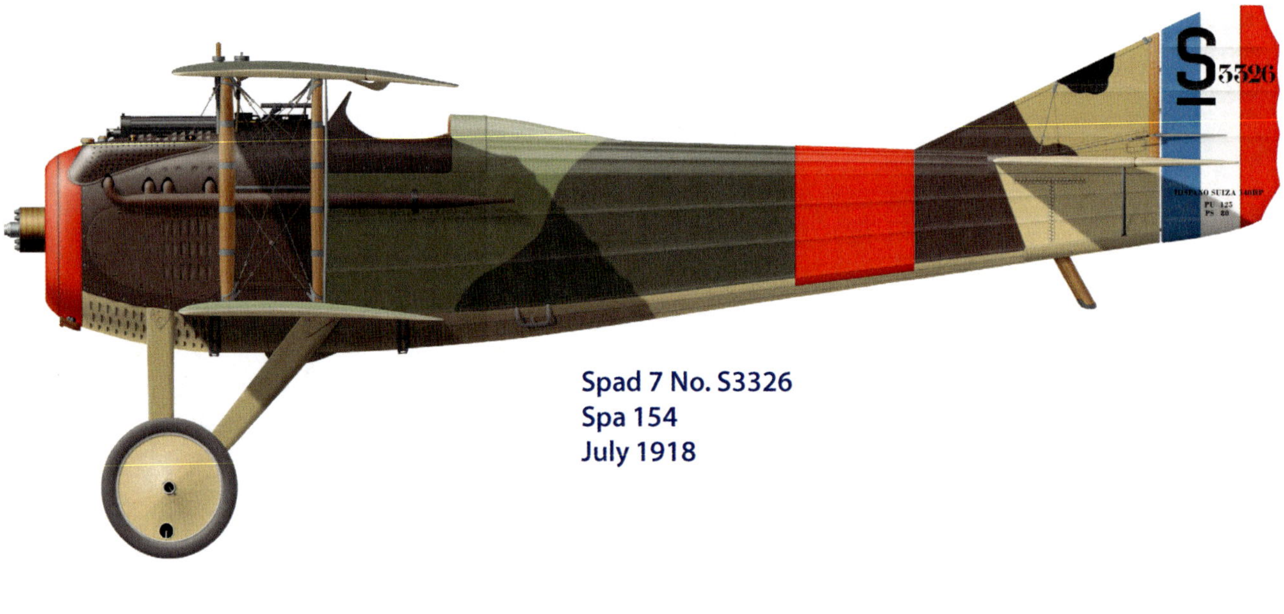

Spad 7 No. S3326
Spa 154
July 1918

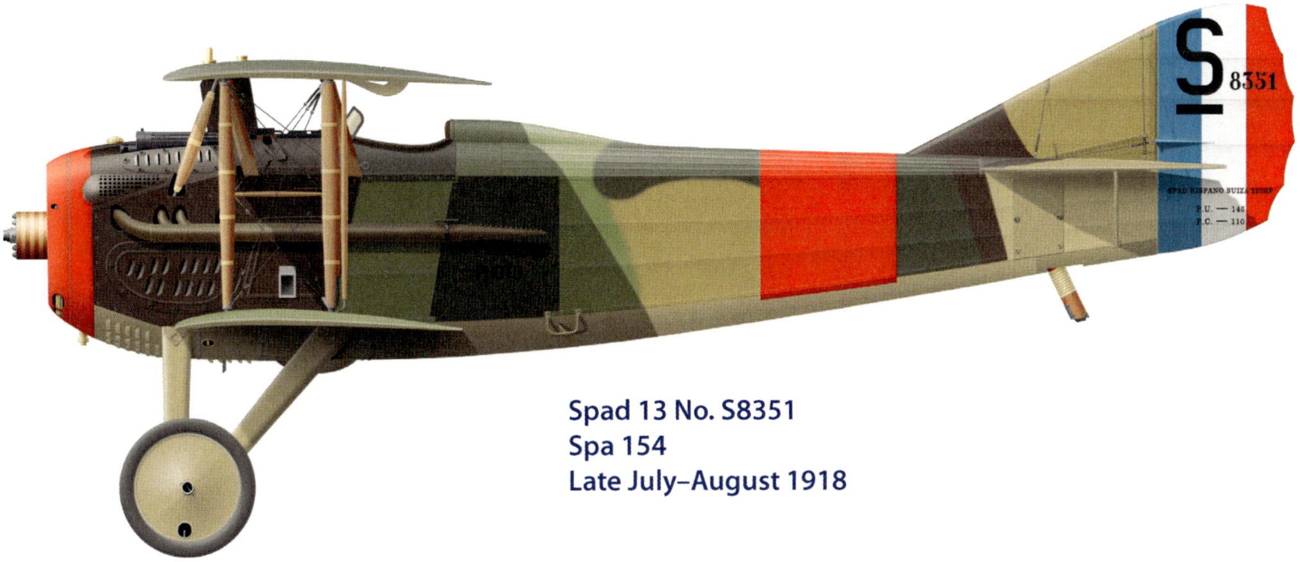

Spad 13 No. S8351
Spa 154
Late July–August 1918

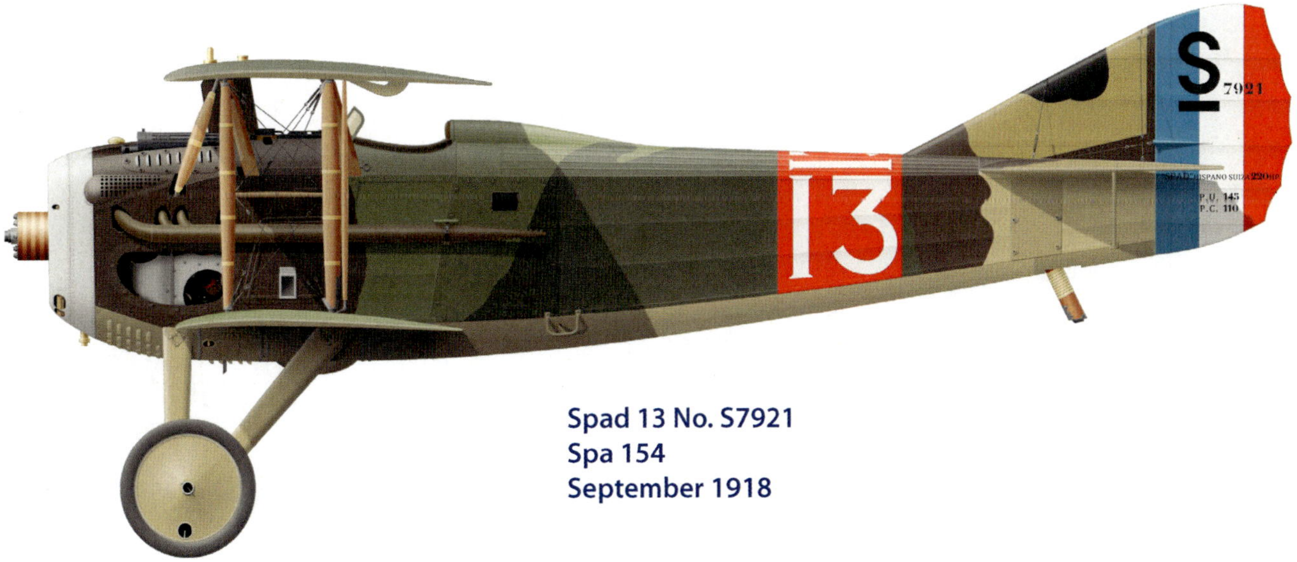

Spad 13 No. S7921
Spa 154
September 1918

Victor Georgievich Fedorov (Victor Fedoroff) (1885-1922)

Lieutenant

C 42, N 26, 11-th KAO, Spa 89

5 victories

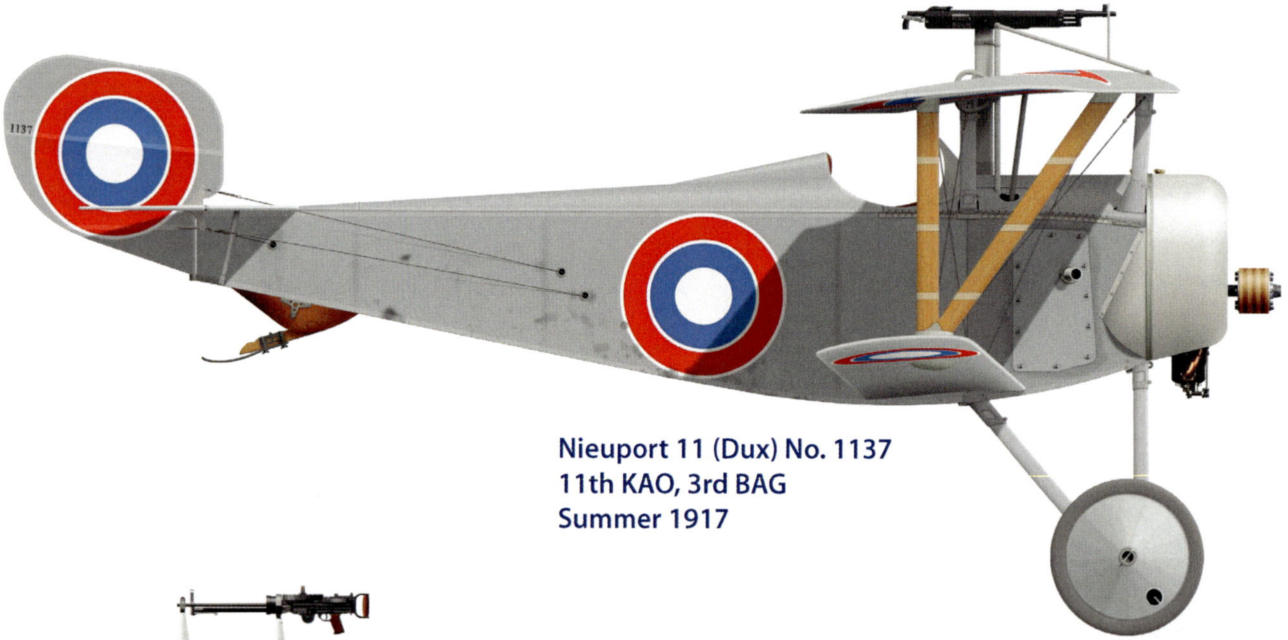

Nieuport 11 (Dux) No. 1137
11th KAO, 3rd BAG
Summer 1917

Nieuport 17 No. 2752
11th KAO, 3rd BAG
Summer 1917

Georges Charles Marie François Flachaire (1892–1973)

Sous-Lieutenant

N 67, N 105

8 victories

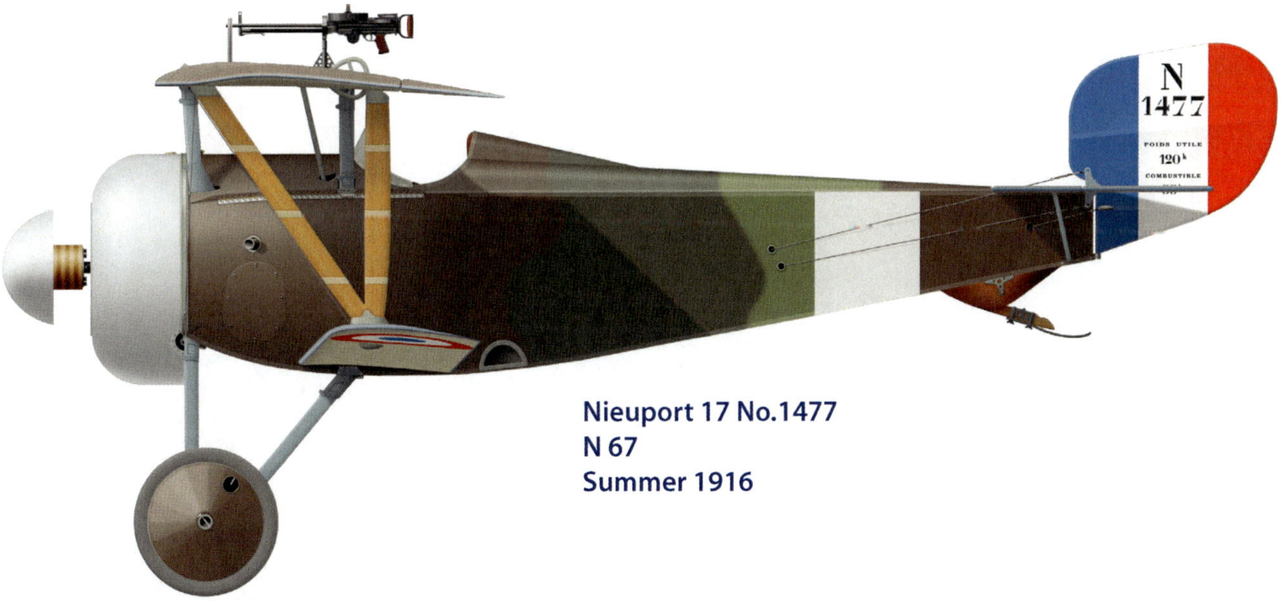

Nieuport 17 No.1477
N 67
Summer 1916

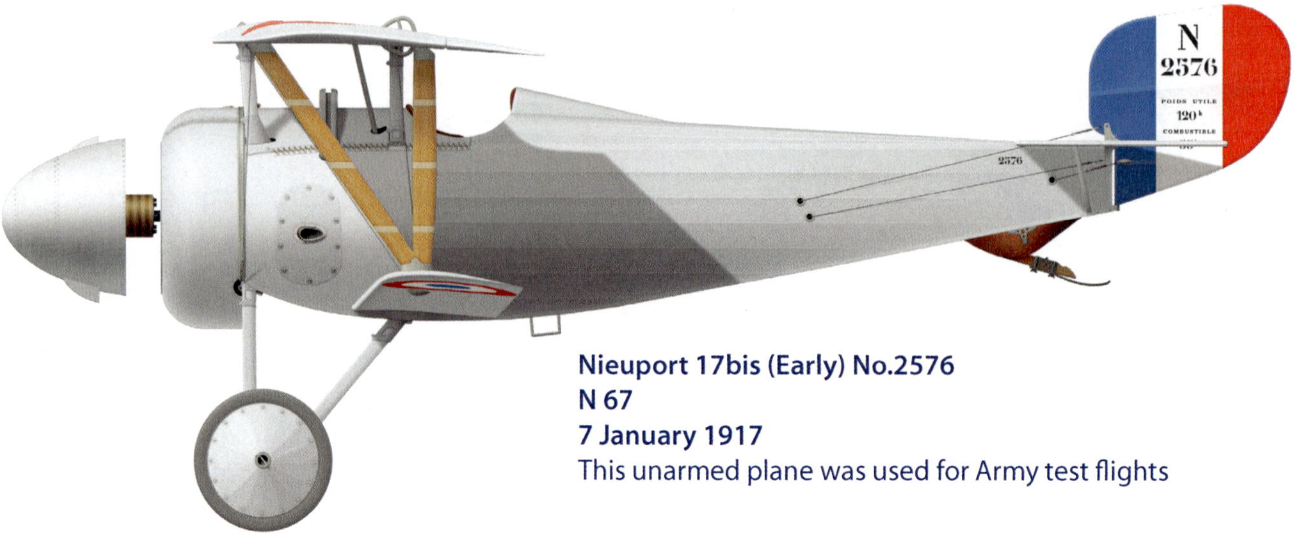

Nieuport 17bis (Early) No.2576
N 67
7 January 1917
This unarmed plane was used for Army test flights

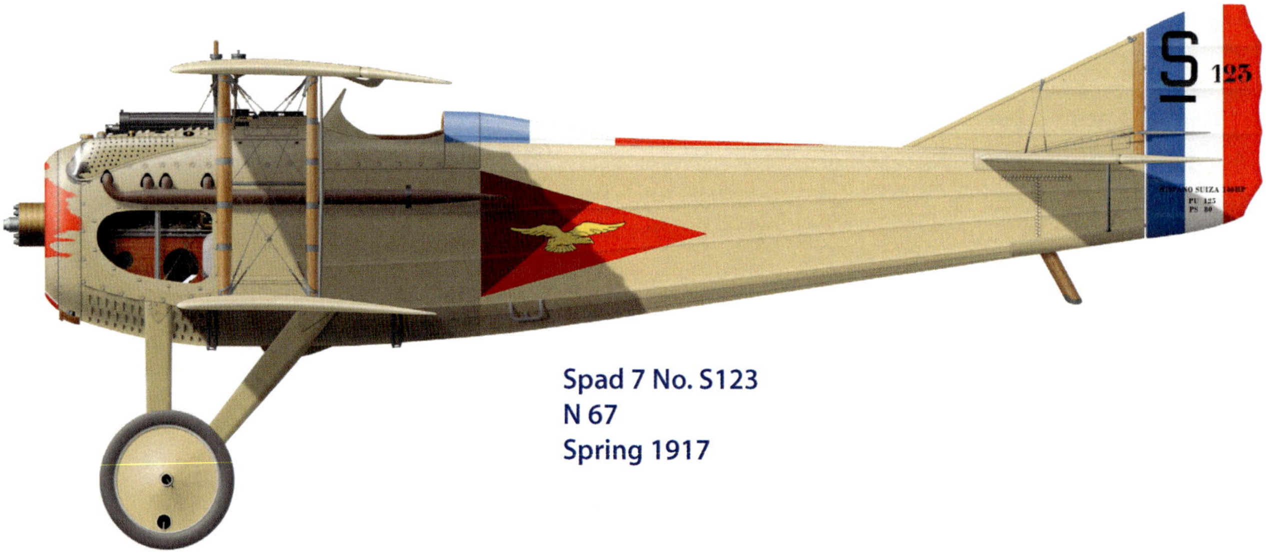

Spad 7 No. S123
N 67
Spring 1917

Spad 13 No. S548
French mission in the US,
Autumn 1917

René Paul Fonck
(1894–1953)
Capitaine
C 47, Spa 103
75 victories

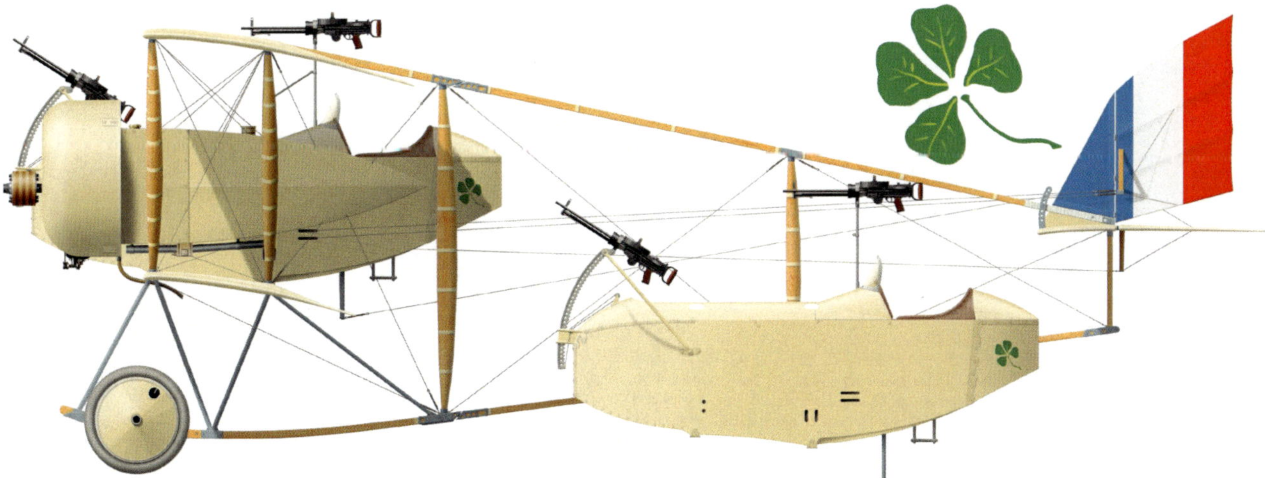

Caudron G.4
C 47
October 1915

Spad 7
Spa 103
Summer 1917

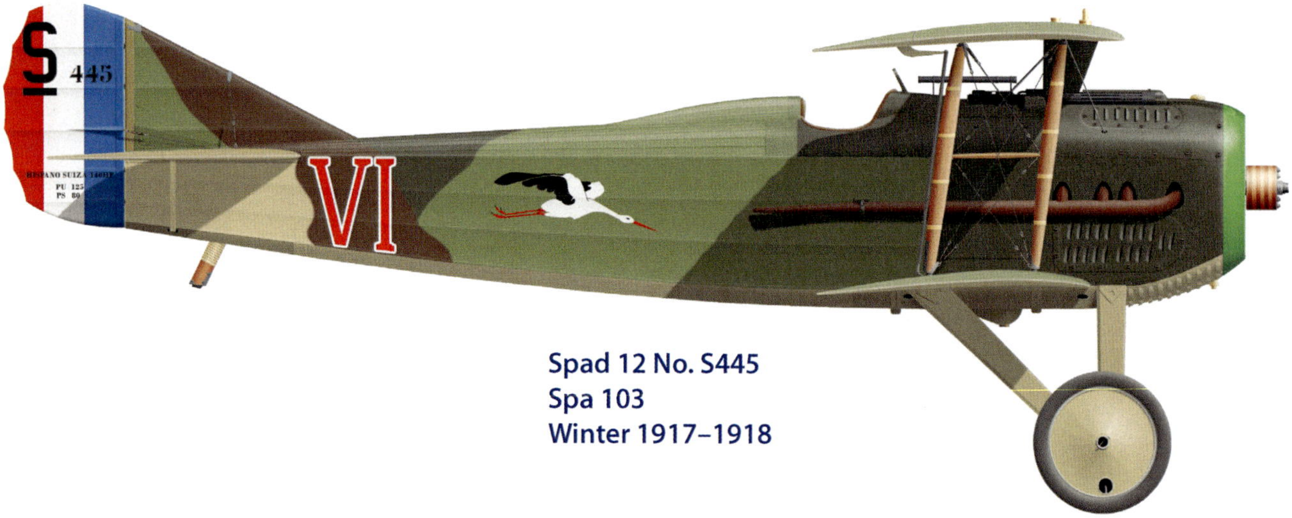

Spad 12 No. S445
Spa 103
Winter 1917–1918

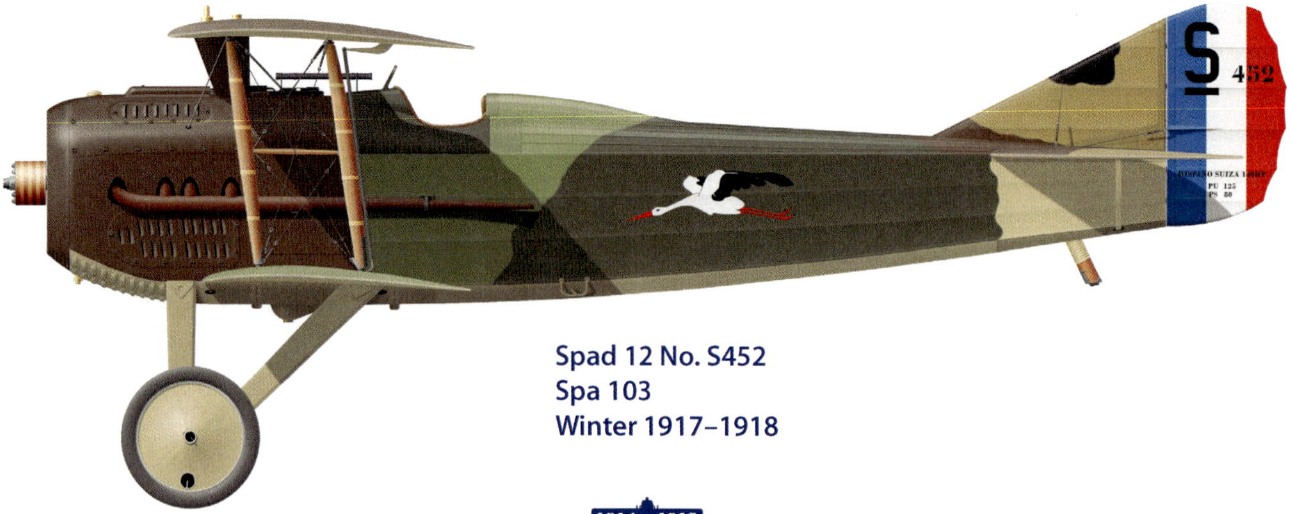

Spad 12 No. S452
Spa 103
Winter 1917–1918

Spad 13 No.S532
Spa 103
February 1918

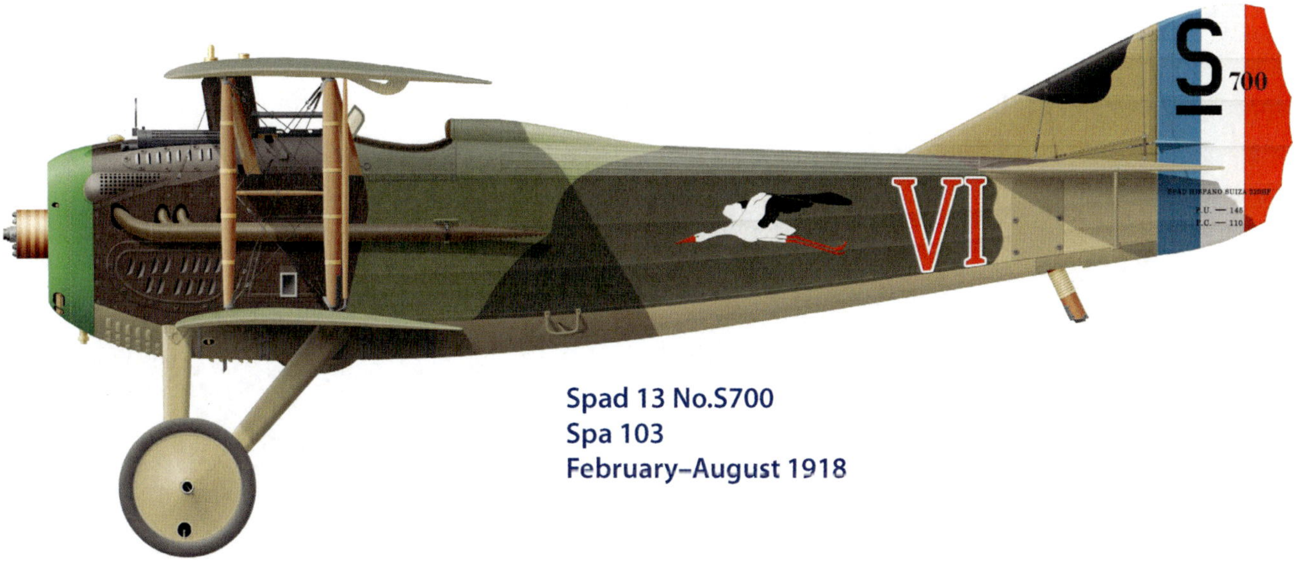

Spad 13 No.S700
Spa 103
February–August 1918

Spad 17 No.S682
Spa 103
October 1918

Jean Alfred Fraissinet
(1894–1981)
Lieutenant
N 57, Spa 57
8 victories

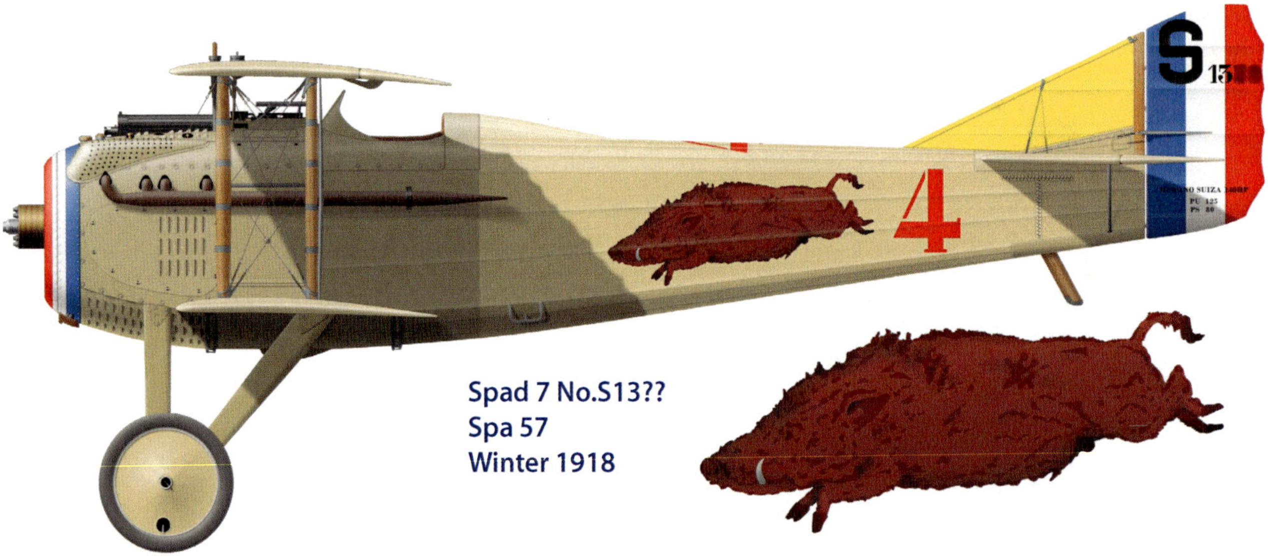

Spad 7 No.S13??
Spa 57
Winter 1918

Morane-Saulnier MS 406C1 No.720 (L-749)
GC II/3
March 1940

Robert René Franc
(1891–1923)
Capitaine
N62, N79, Spa79
5 victories

Spad 7
Spa 79
Autumn 1917

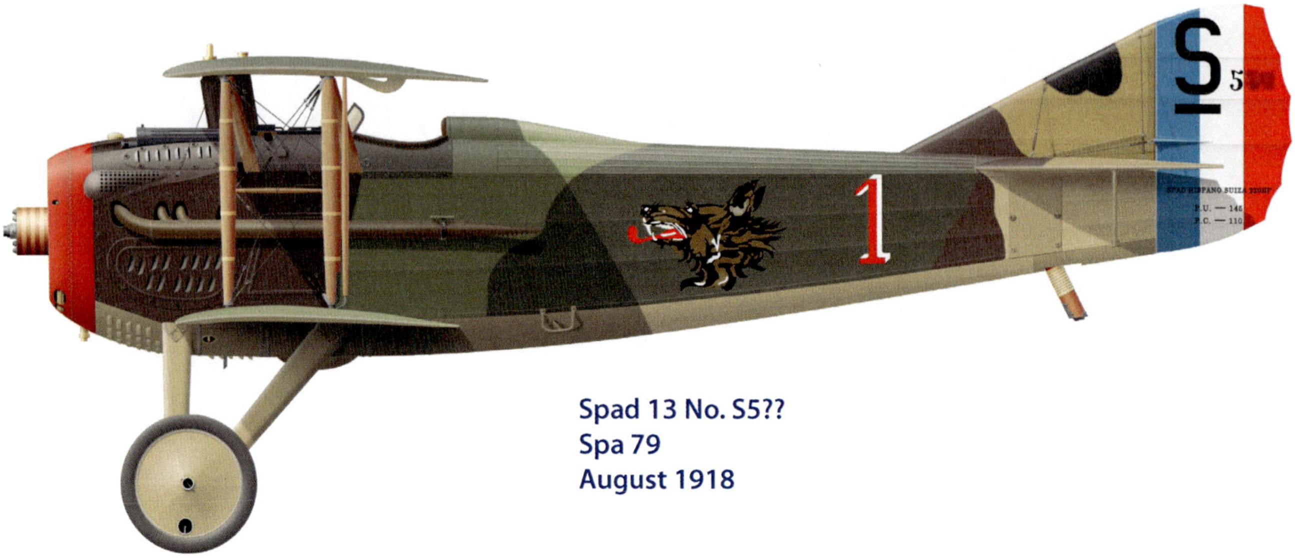

Spad 13 No. S5??
Spa 79
August 1918

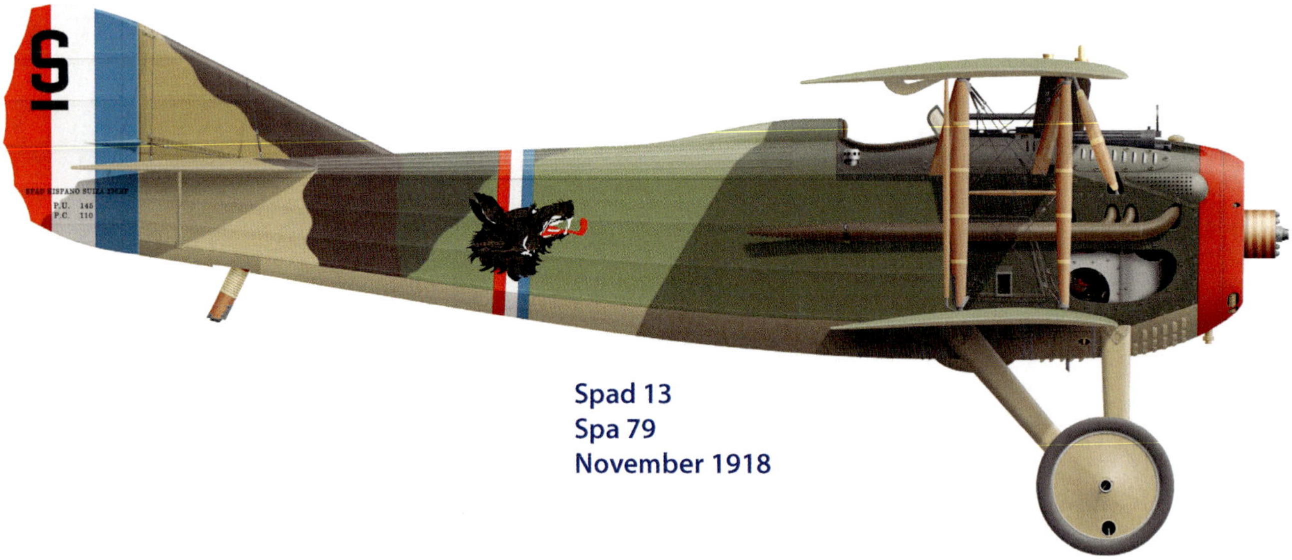

Spad 13
Spa 79
November 1918

Jean Pierre Marie Joseph de Gaillard de la Valden (1895-1977)

Maréchal-de-Logis
MF123, Spa95
5 victories

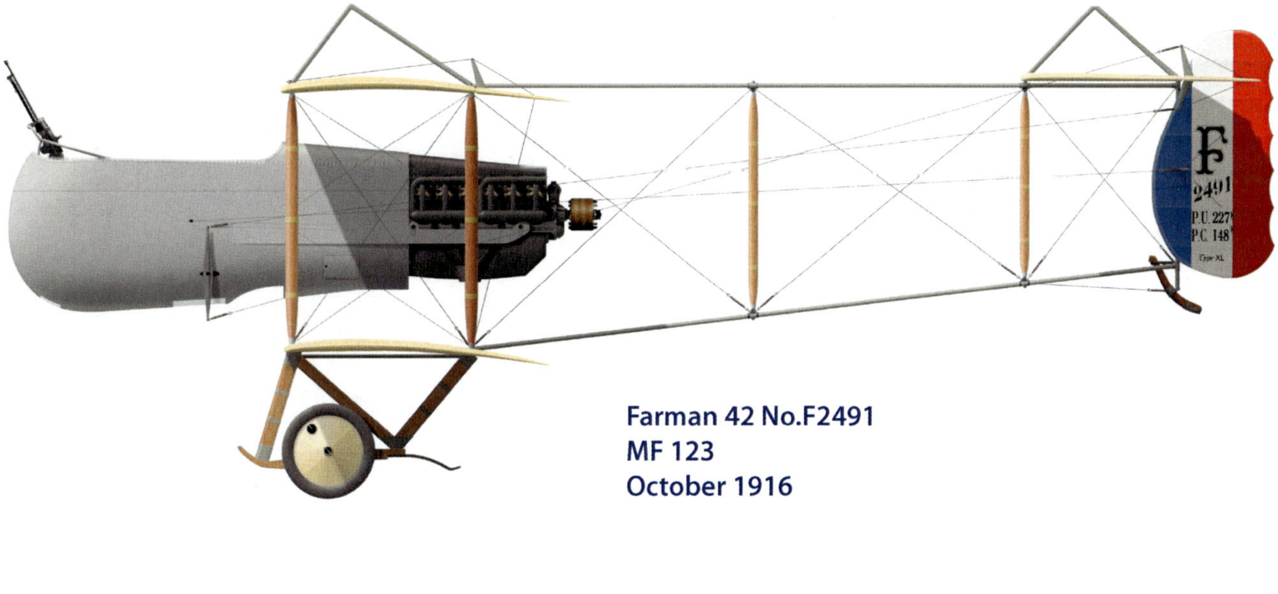

Farman 42 No.F2491
MF 123
October 1916

Hector Eugène Joseph Garaud
(1897–1940)
Sous-Lieutenant
V 97, Spa 38
13 victories

Voisin LAS 3 No. V.1535
Esc V 97 CPR
Spring 1916

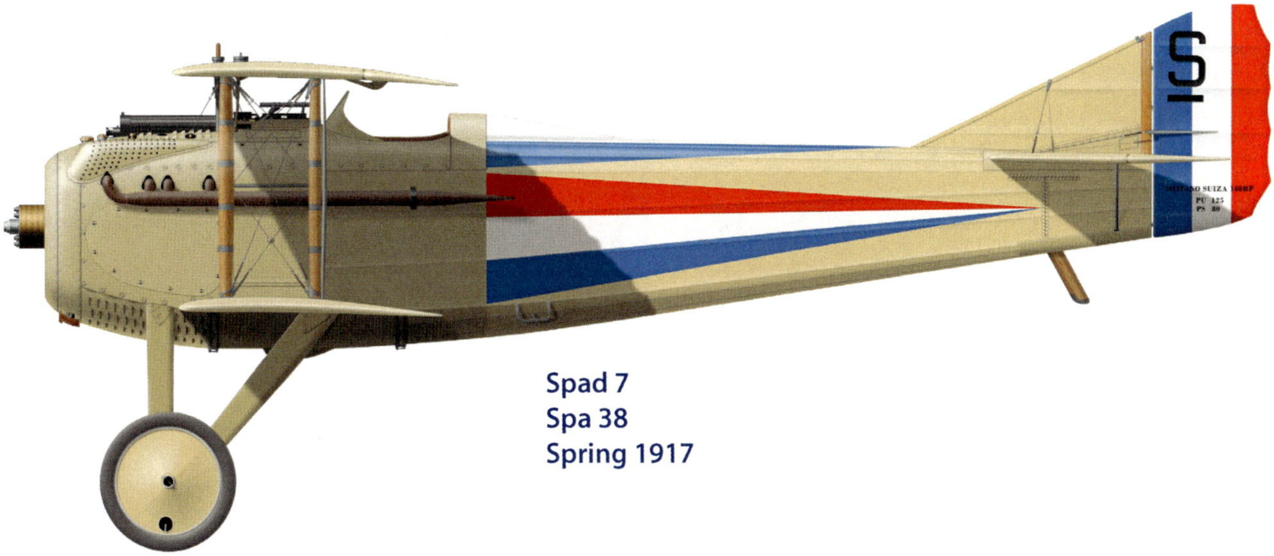

Spad 7
Spa 38
Spring 1917

Spad 7 No. S1165
Spa 38
Summer 1917

Lucien Marcel Gasser

(1897–1939)
Adjudant
Spa 87
10 victories

Nieuport 17
Flying School Avord
Autumn 1917

Nieuport 27 No. N5690
N 87
Early 1918

Paul Adrien Gastin
(1886–1976)
Capitaine
MS12, N49, N84, Spa84, Group de Combat 23
6 victories

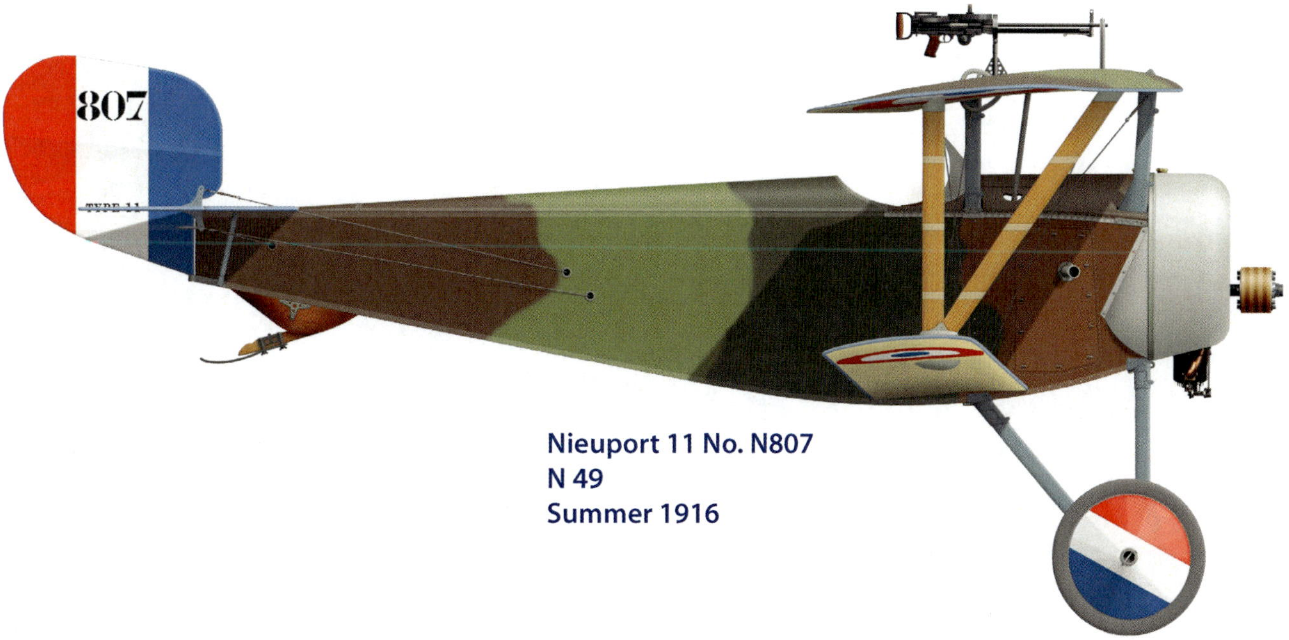

Nieuport 11 No. N807
N 49
Summer 1916

Pierre Gustave Gaudermen
(1882–1948)
Sous-Lieutenant
MF 55, N 68, V 110, V 114, Spa 68
5 victories

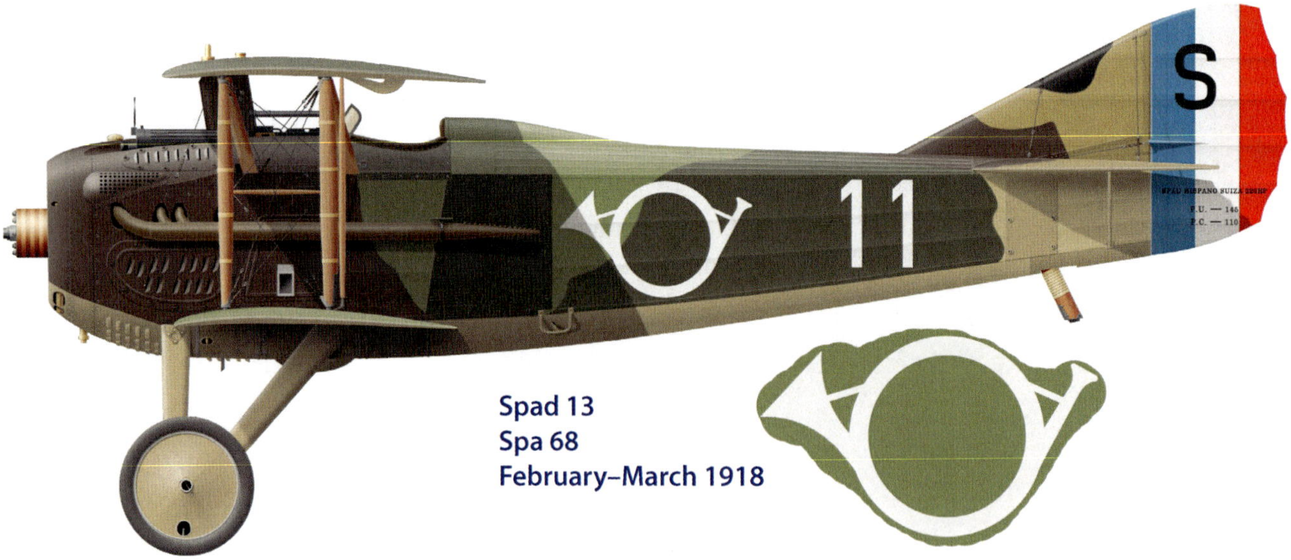

Spad 13
Spa 68
February–March 1918

Alexander Emile Jacques Gerard

(1890–1918)
Adjudant
C 18, Spa 65
8 victories

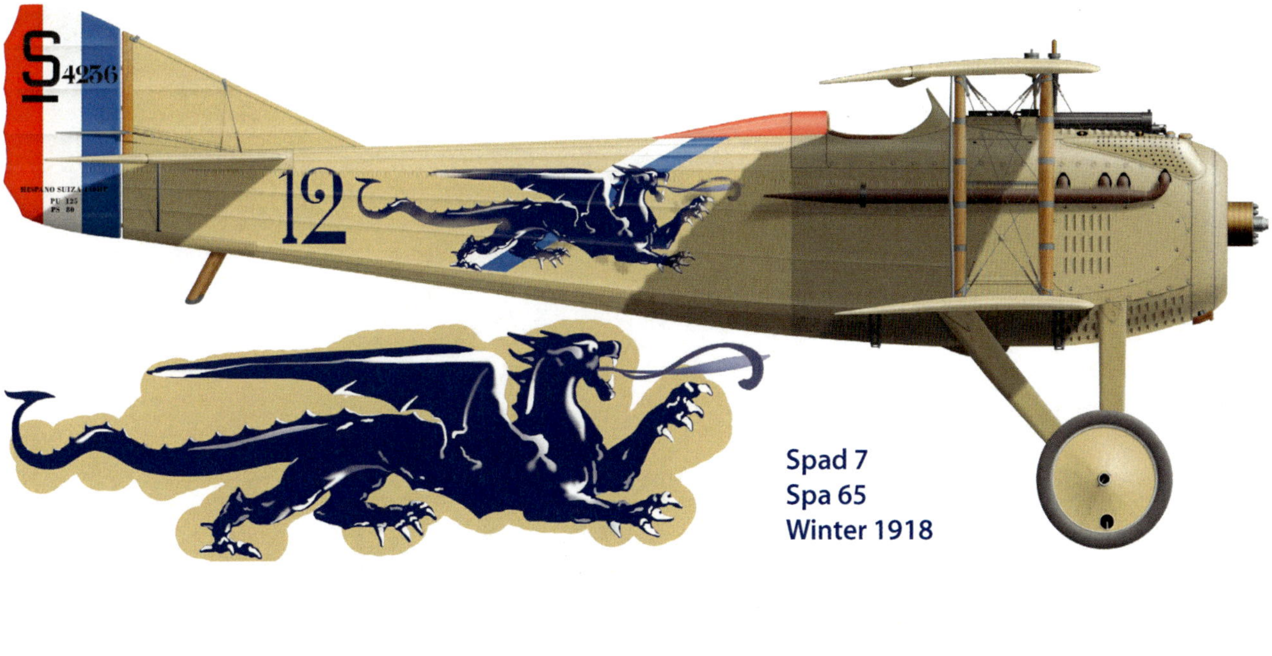

Spad 7
Spa 65
Winter 1918

Adriene-Eugène Gilbert
(1889–1918)
Sous-Lieutenant
MS 23, MS 37, MS 49
5 victories

Morane-Saulnier L
MS 23
Autumn 1914

Morane-Saulnier N *LE VENGEUR* No. MS388
MS 49
June 1915

Morane-Saulnier L No. MS245
MS 49
27 June 1915

Morane-Saulnier AI No. MS3
17 May 1918
The day Gilbert was killed in the crash of this airplane

Maurice Roch Gond

(1884–1964)
Capitaine
N 67, N 3 (Franco-Romanian)
6 victories

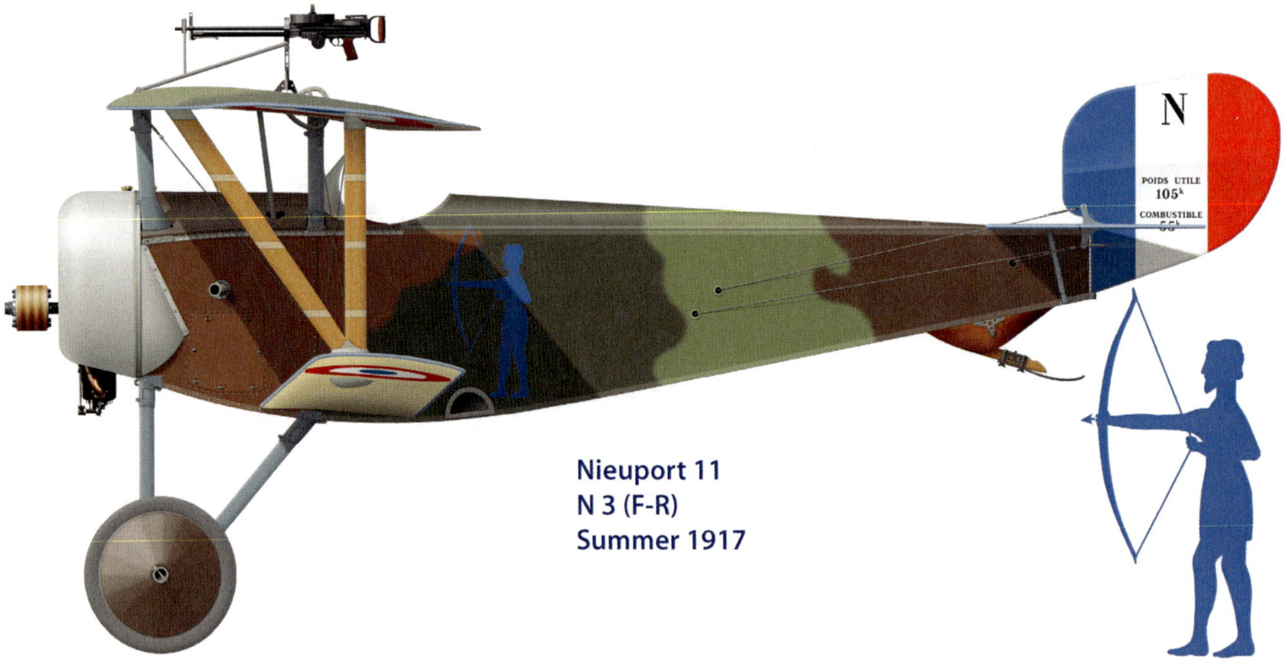

Nieuport 11
N 3 (F-R)
Summer 1917

Nieuport 11 No.1405
N 3 (F-R)
Summer 1917

Nieuport 21 No.1909
N 3 (F-R)
July 1917
Reconstruction

Louis Prosper Gros
(1893–1973)
Sous-Lieutenant
HF 41, Spa 154
9 victories

Spad 13 No. S2215
Spa 154
April 1918

Gabriel Fernand Charles Guérin
(1892–1918)
Lieutenant
Spa 15, Spa 88
23 victories

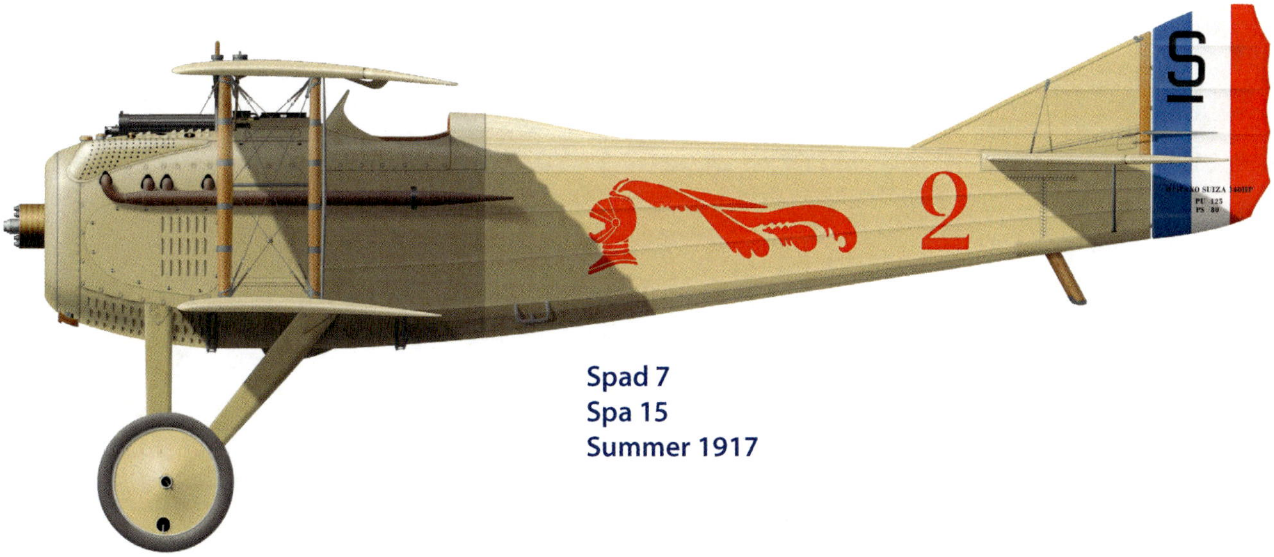

Spad 7
Spa 15
Summer 1917

Spad 12 No. S444
Spa 15
Winter 1918

Julien Anatole Guertiau
(1885–1954)
Lieutenant
C43, Spa94
8 victories

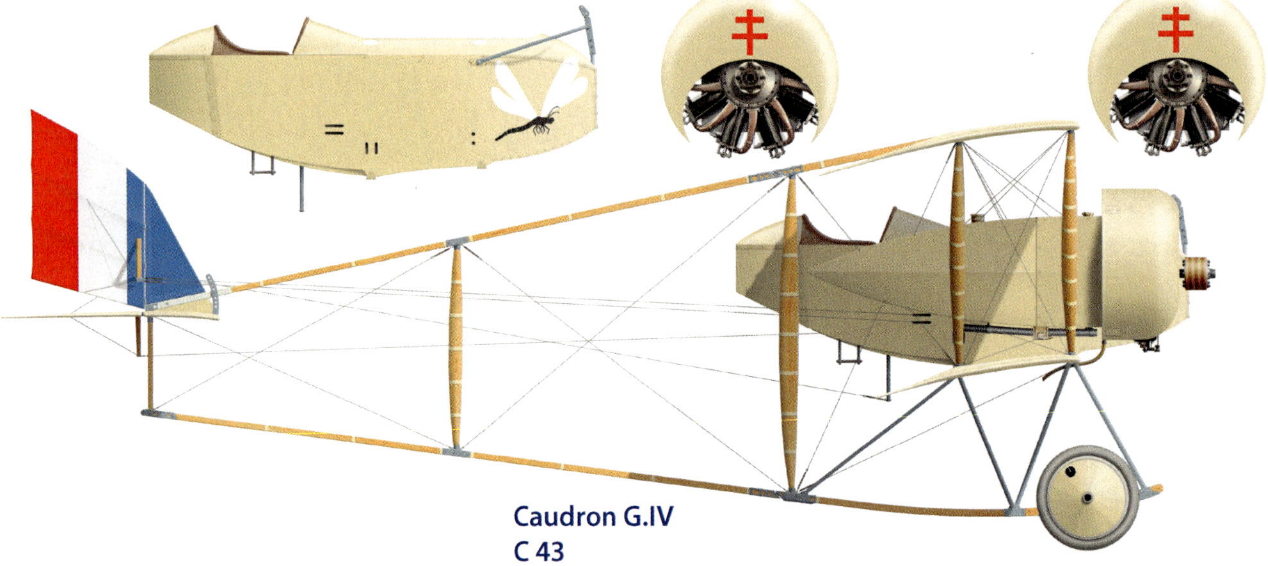

Caudron G.IV
C 43
October 1916

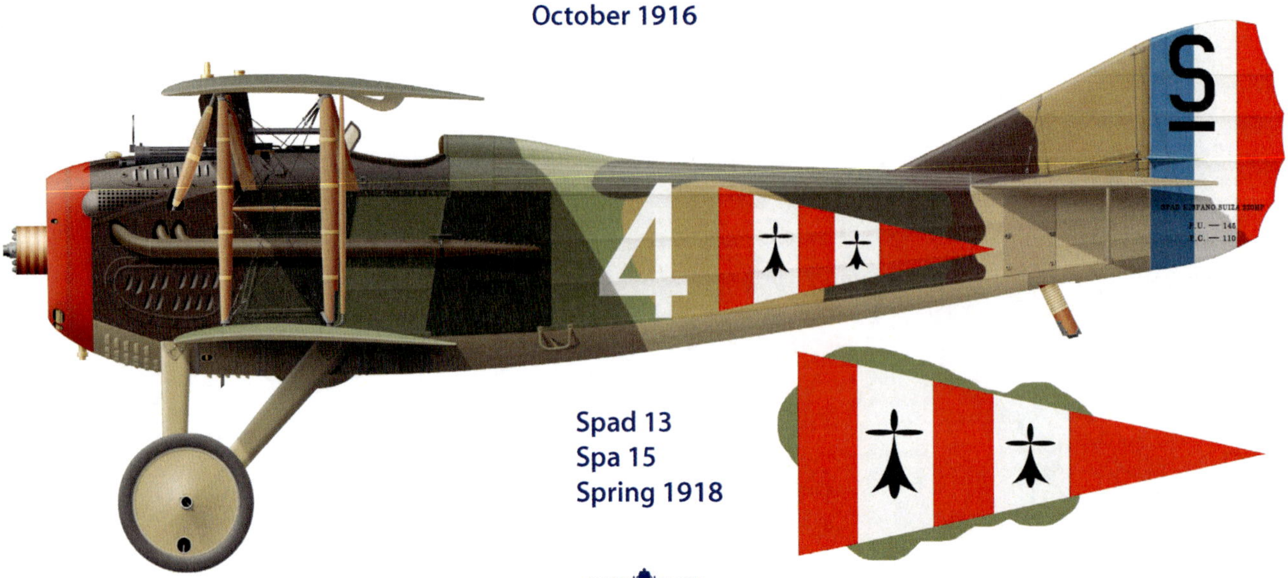

Spad 13
Spa 15
Spring 1918

Joseph-Henri Guiguet
(1891–1979)
Sous-Lieutenant
HF 1, C 94 CRP, N 3, N 95, Spa 167
5 victories

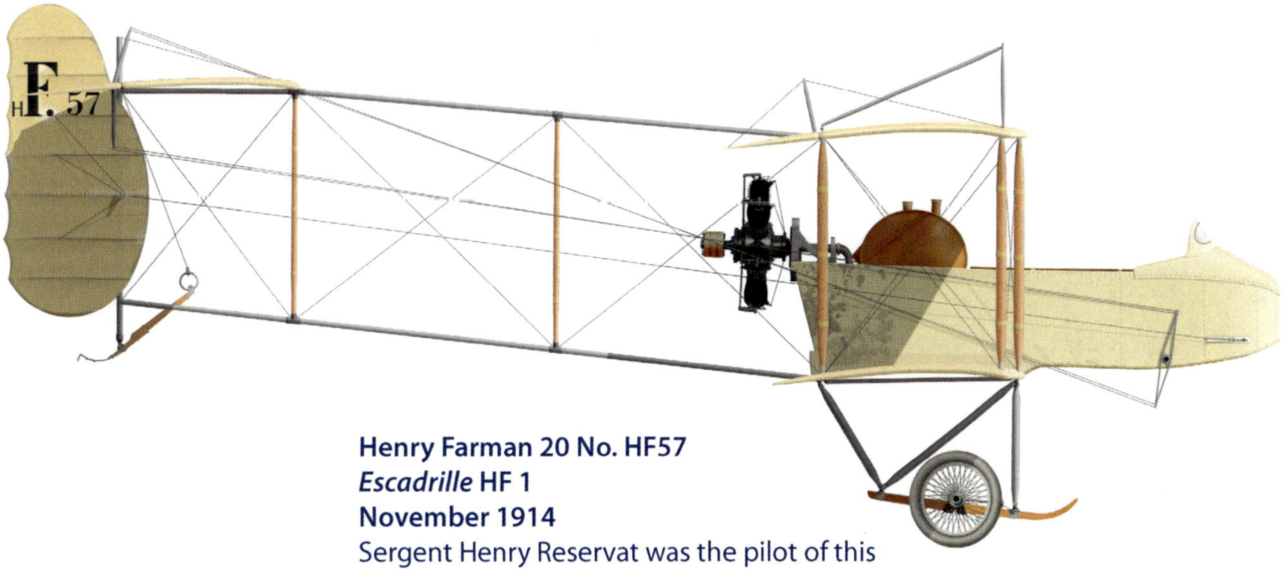

Henry Farman 20 No. HF57
***Escadrille* HF 1**
November 1914
Sergent Henry Reservat was the pilot of this plane; Guiguet was the observer.

Blériot XI-2 No. BL182
Aviation School Pau
May 1915

Caudron G.3 No. C1003
C 94 CPR
June 1915

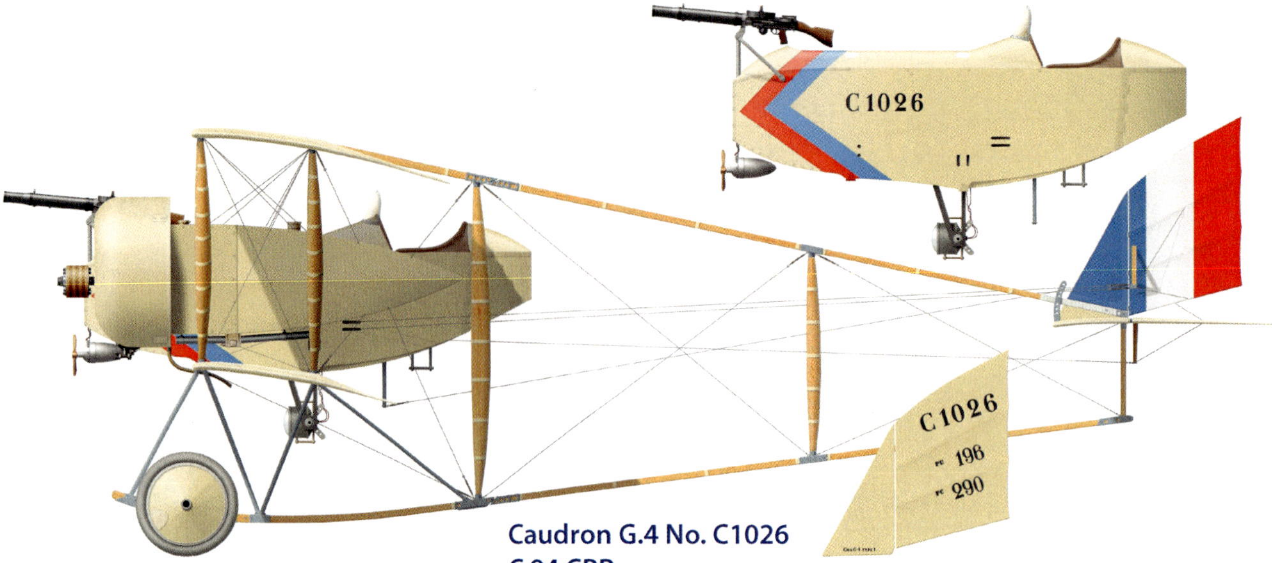

Caudron G.4 No. C1026
C 94 CPR
November 1915

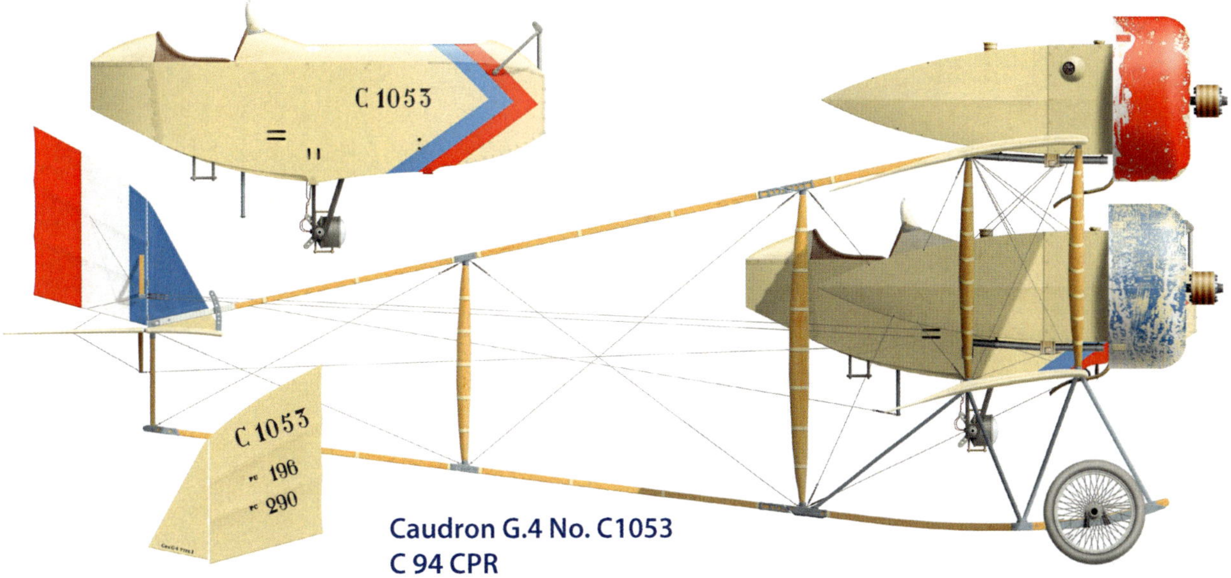

Caudron G.4 No. C1053
C 94 CPR
Winter 1915–1916

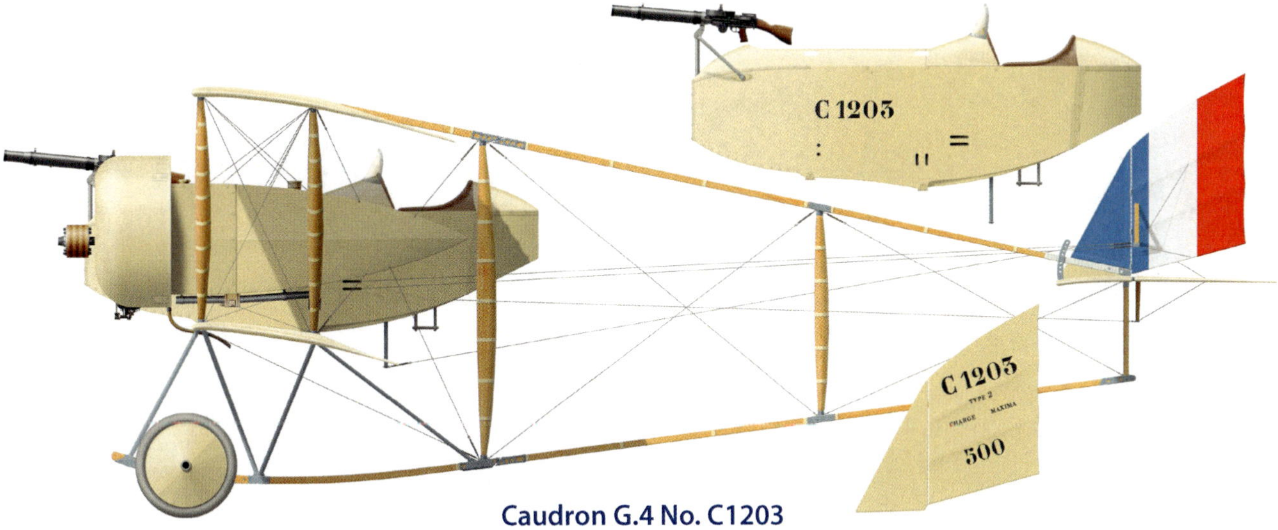

Caudron G.4 No. C1203
C 94 CPR
Winter 1915–1916

Nieuport 10 No. N268
N 95 CPR
March 1916

Nieuport 16 No. N976
N 95 CPR
April 1916

Nieuport 16 No. N978
N 95 CPR
May 1916

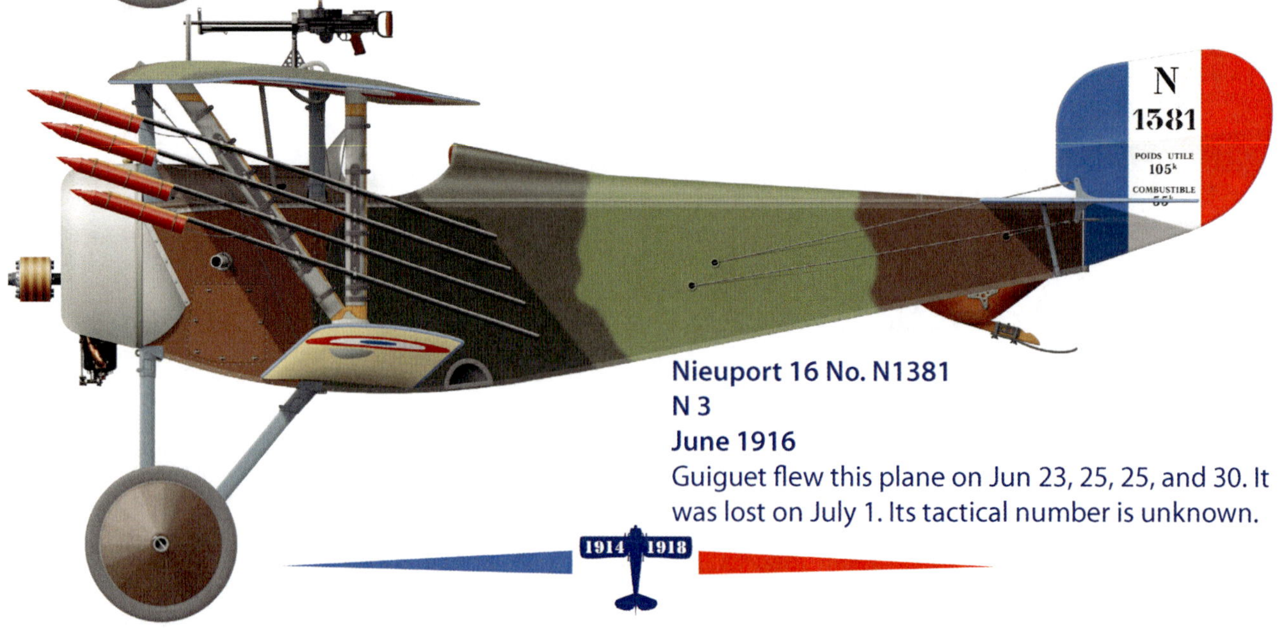

Nieuport 16 No. N1381
N 3
June 1916
Guiguet flew this plane on Jun 23, 25, 25, and 30. It was lost on July 1. Its tactical number is unknown.

Nieuport 16 No. N939
N 3
June 1916
This is Felix Brocard's machine. Guiguet flew it on June 18, 21, and 26.

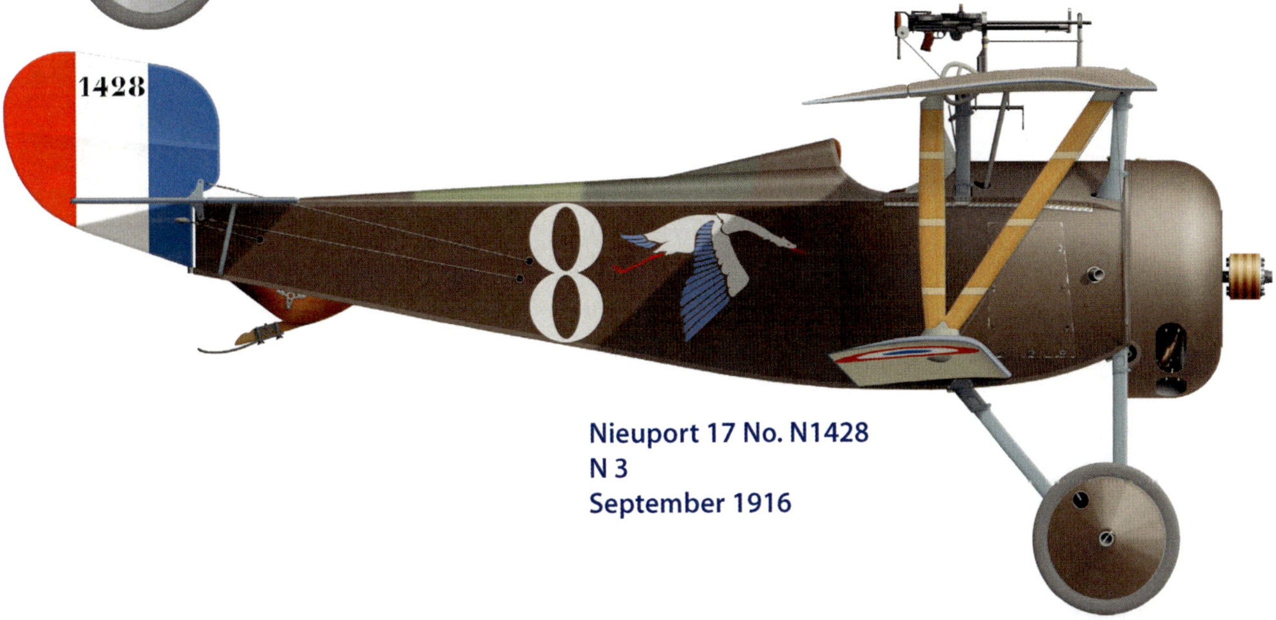

Nieuport 17 No. N1428
N 3
September 1916

Nieuport 17 *Ptit Jo II* No. N2007
N 3
November–December 1916

Spad 7 *Ptit Jo III* No. S1105
N 3
April 1917

Spad 7 *Ptit Jo IV* No. S1418
N 3
May 1917

Spad 13 *Ptit Jo V*
Spa 167
October 1918

1914 1918

Gilbert Marie de Guingand
(1891–1918)
Sous-Lieutenant
C 34, N 15, N 88, N 48
8 victories

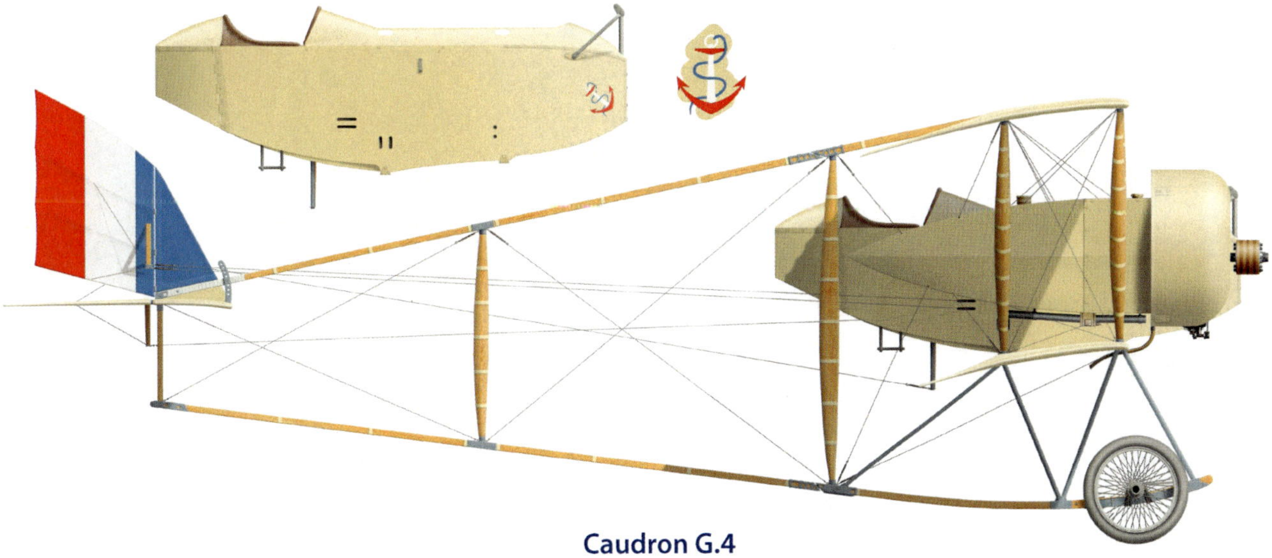

Caudron G.4
C 34
Summer 1916

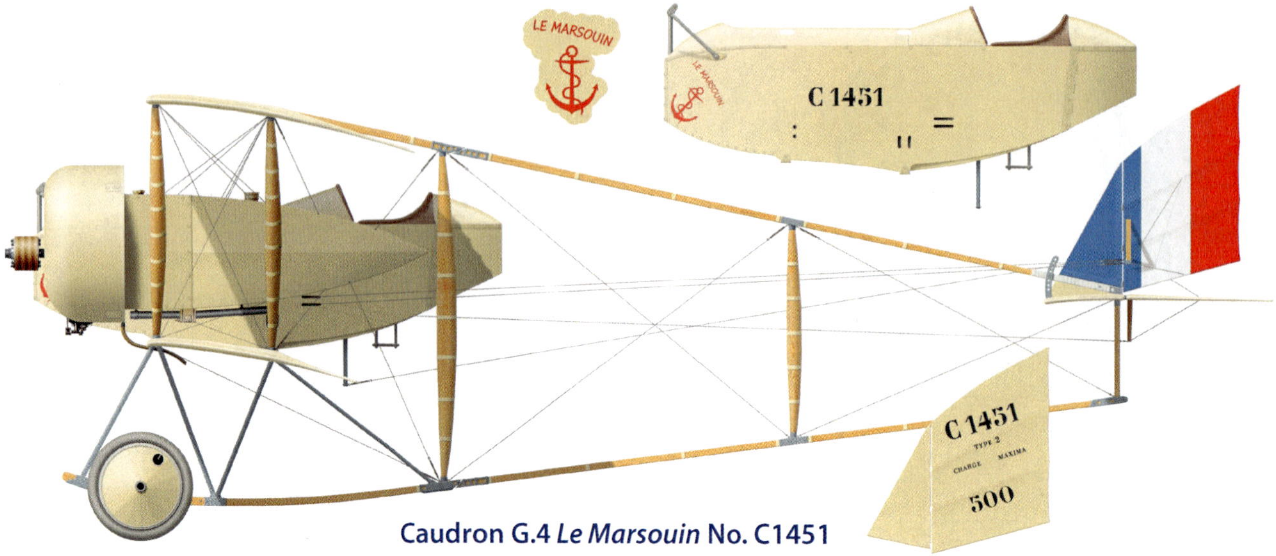

Caudron G.4 *Le Marsouin* No. C1451
C 34
Summer 1916

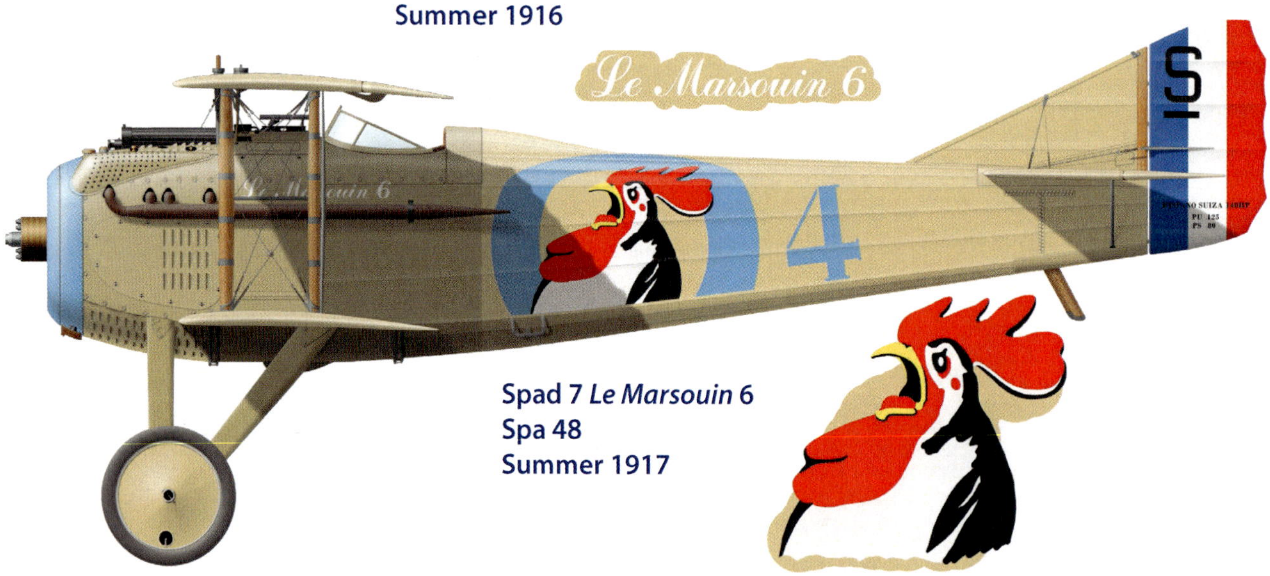

Spad 7 *Le Marsouin* 6
Spa 48
Summer 1917

Spad 13
Spa 48
October 1918

Georges Marie Ludovic Jules Guynemer (1894–1917)

Capitaine

MS 3, N 3, Spa 3

53 victories

Morane-Saulnier H
Flying School Avord
April 1915

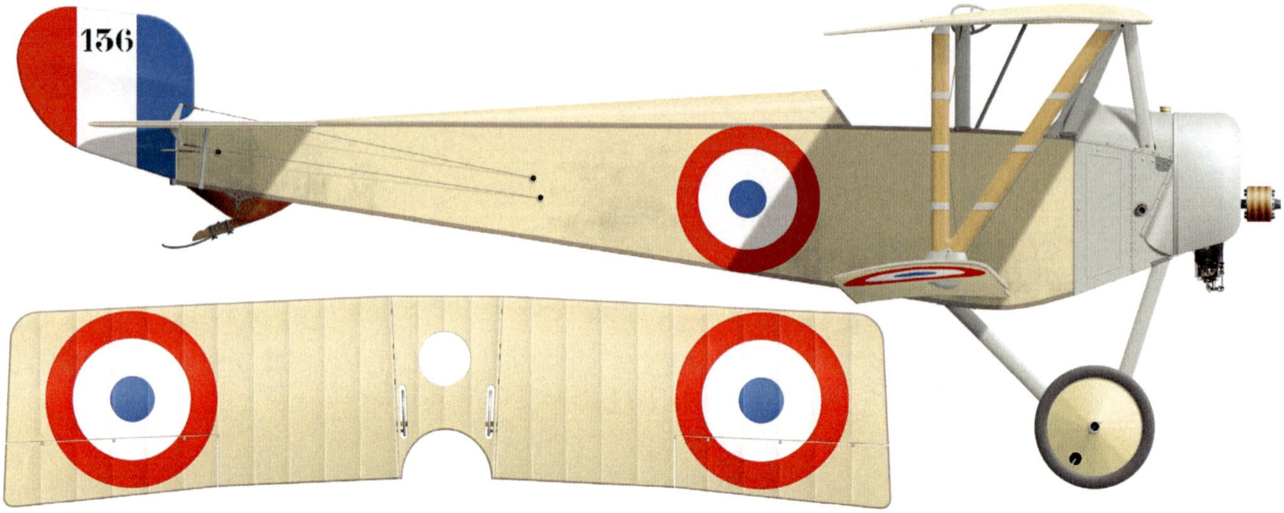

Nieuport 10B No. N136
MS 3
June 1915

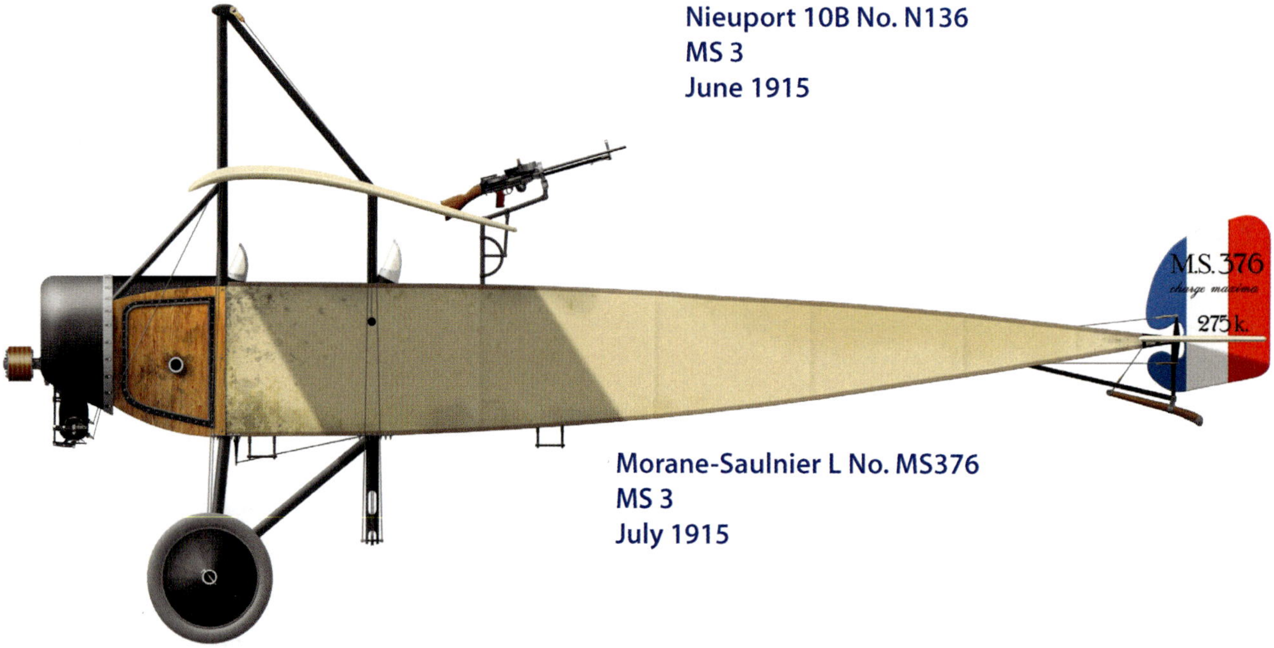

Morane-Saulnier L No. MS376
MS 3
July 1915

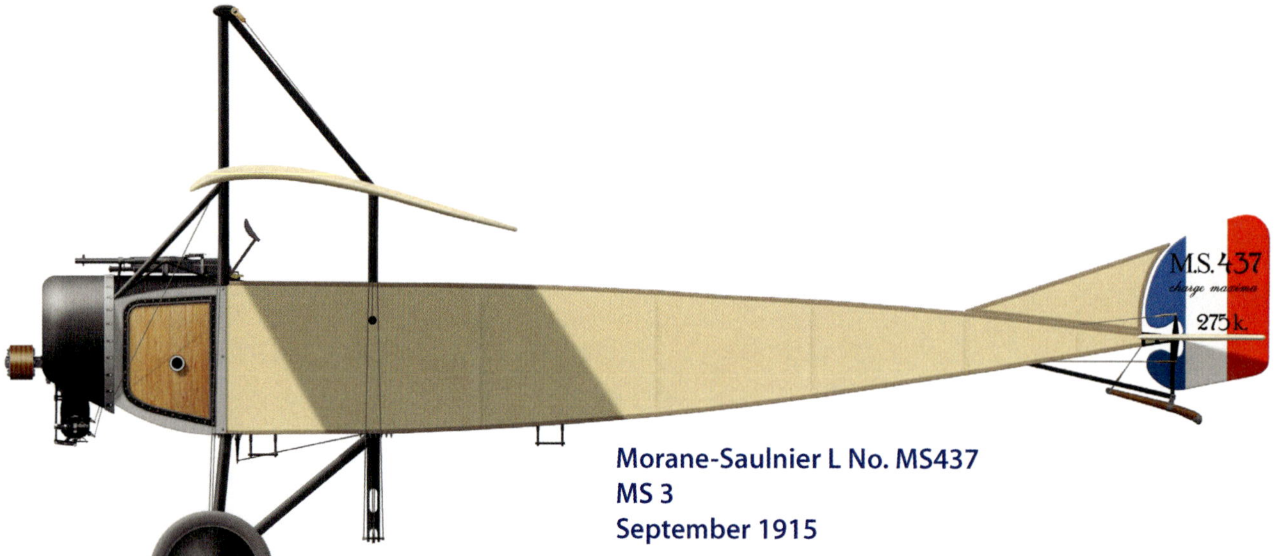

Morane-Saulnier L No. MS437
MS 3
September 1915

Nieuport 10 No. N302
MS 3
September 1915–March 1916

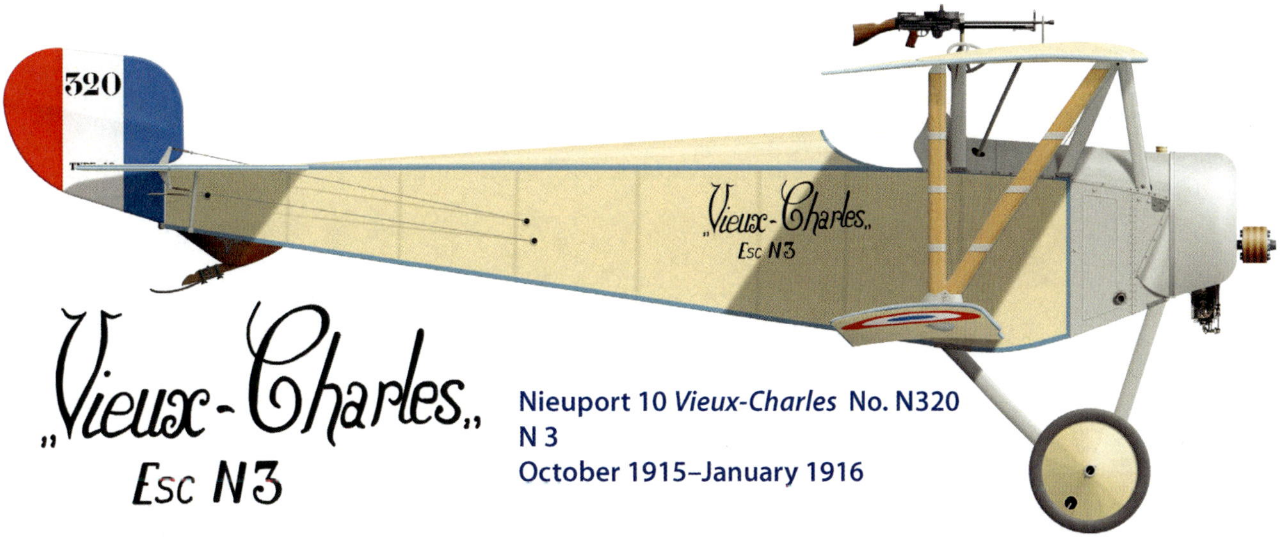

Nieuport 10 *Vieux-Charles* No. N320
N 3
October 1915–January 1916

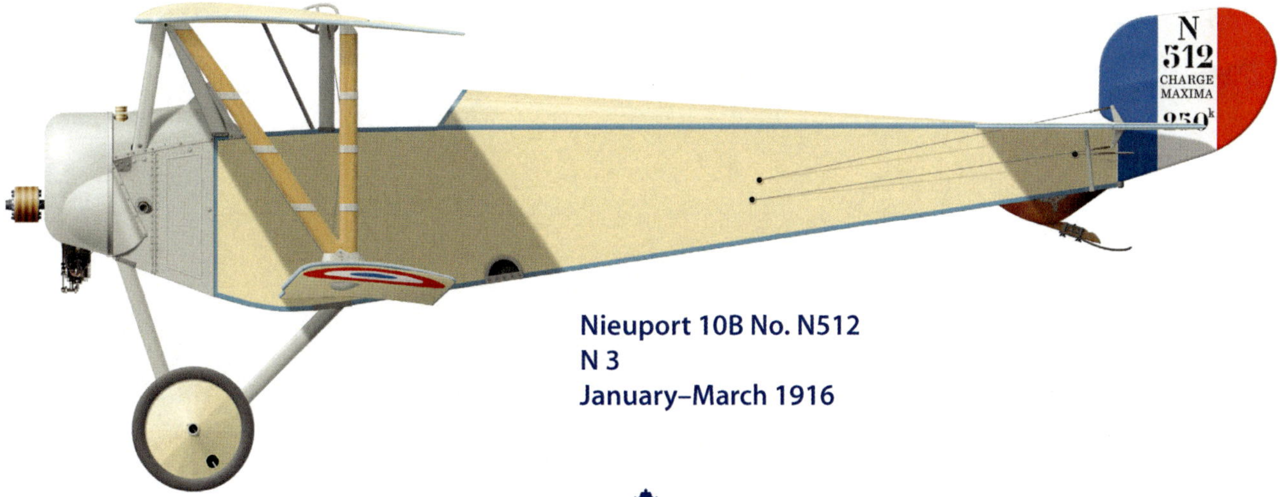

Nieuport 10B No. N512
N 3
January–March 1916

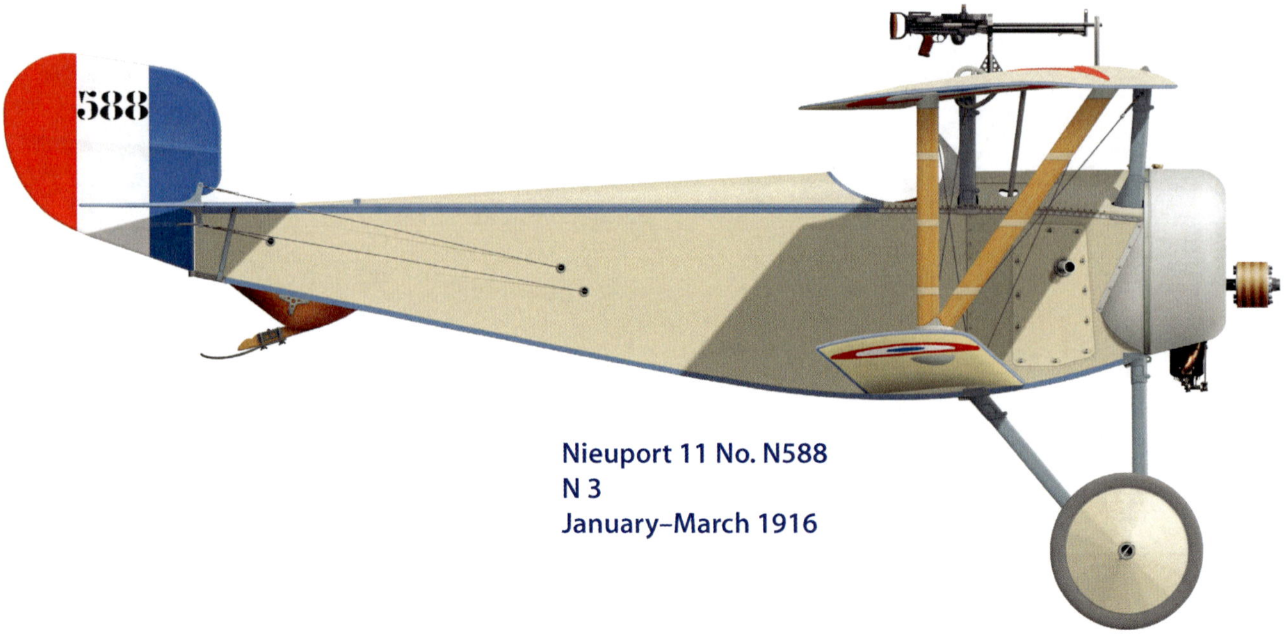

Nieuport 11 No. N588
N 3
January–March 1916

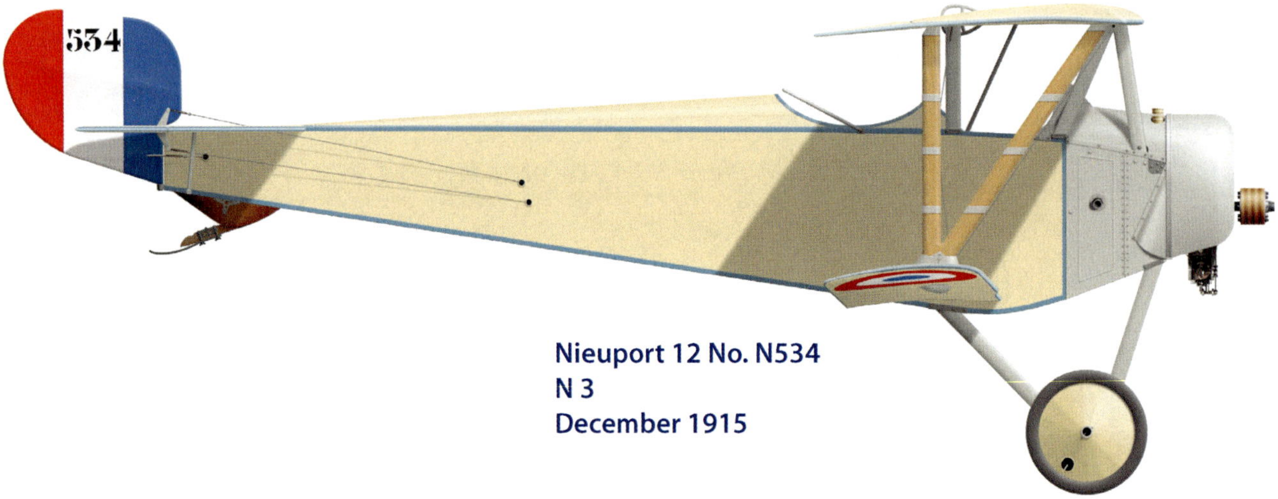

Nieuport 12 No. N534
N 3
December 1915

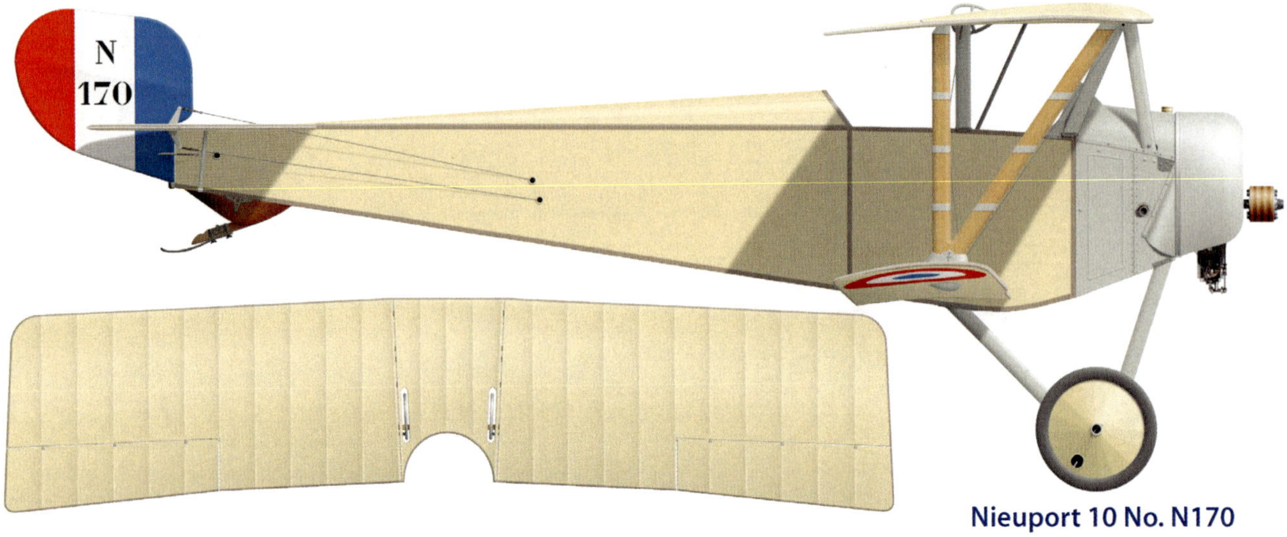

Nieuport 10 No. N170
N 3
December 1915

Nieuport 11 *Vieux-Charles* No. N836
N 3
March 1916

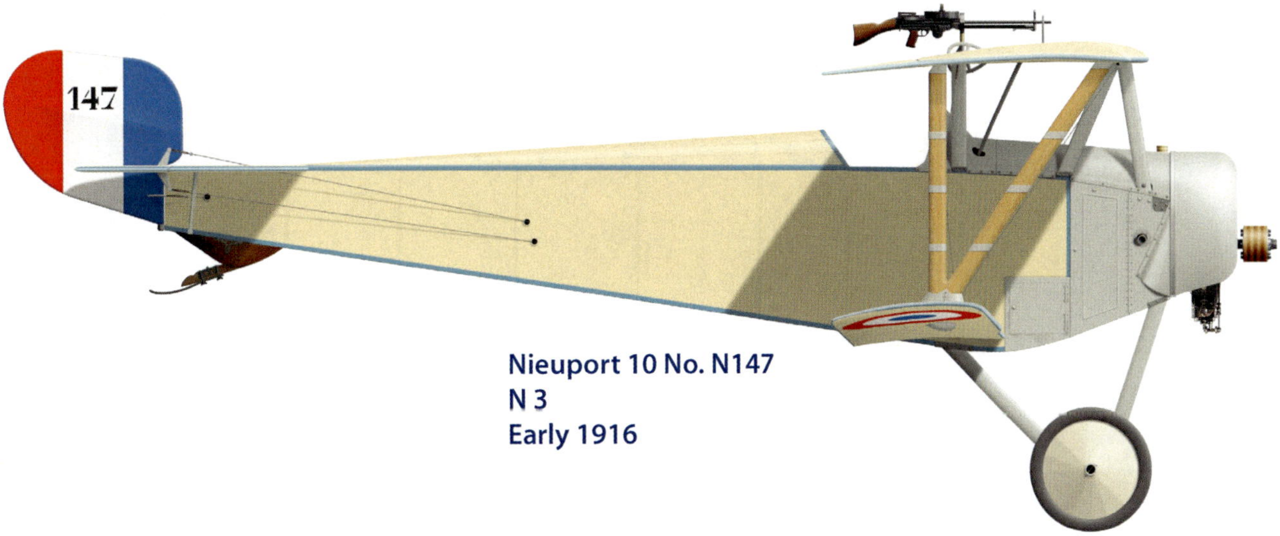

Nieuport 10 No. N147
N 3
Early 1916

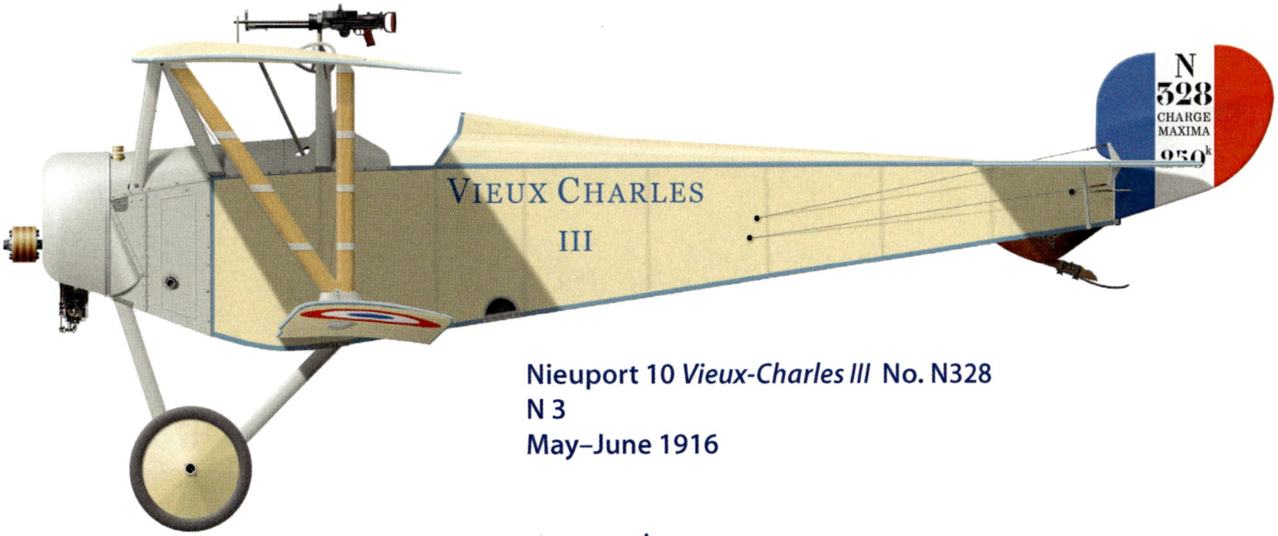

Nieuport 10 *Vieux-Charles III* No. N328
N 3
May–June 1916

Nieuport 16 No. N1386
N 3
July 1916

Nieuport 17 *Vieux-Charles* No. N1531
N 3
July–August 1916

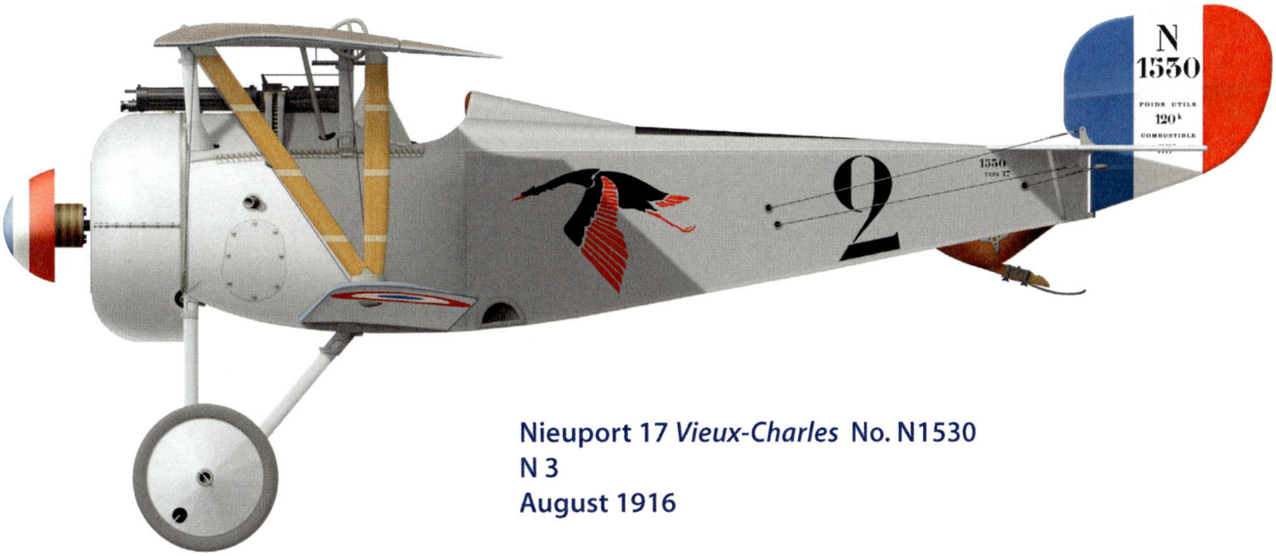

Nieuport 17 *Vieux-Charles* No. N1530
N 3
August 1916

Spad 7 *Vieux-Charles* No. S115
N 3
September 1916

Spad 7 *Vieux-Charles* No. S132
N 3
November 1916

Spad 7 *Vieux-Charles* No. S254
N 3
January–May 1917

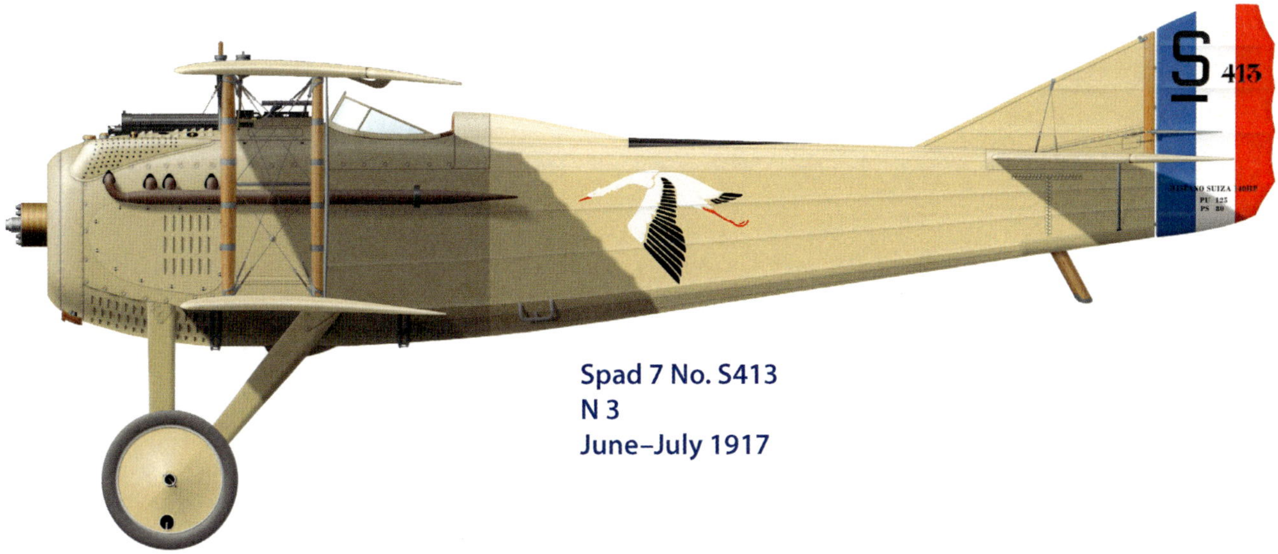

Spad 7 No. S413
N 3
June–July 1917

Spad 12 *Vieux-Charles* No. S382
Spa 3
July–August 1917

Spad 13 No. S504
Spa 3
August–September 1917

Sources & References

Books

Ezra Bowen, *Knights of the Air*, Time-Life Books, 1980
Marc Chassard, Nieuport 17 Analyse détaillée des premières séries, Association FAREWO, 2018.
Marc Chassard, *René Dorme et Joseph Guiguet. La Guerre aérienne de deux As*, ÉDITIONS AUX ARTS, 2002.
James J. Davilla, Artur M. Soltan, *French Aircraft of the First World War*, Flying Machines Press, 1997.
Alan Durkota, Thomas Darcey, Victor Kulikov, *The Imperial Russian Air Service. Famous Pilots and Aircraft of World War I*, Flying Machines Press, 1995.
A. van Haute, *Pictorial History of the French Air Force*, Ian Allan, London, 1974.
Jack Herris, *SPAD Two-seat Fighters of World War I*, Flying Machines Press, 2005.
Stephen Longstreet, *The Canvas Falcons*, Barnes & Noble, 1970
Norman L.R. Franks, Frank W. Bailey, *Over the Front*, Grub Street, 1992.
Norman L.R. Franks, Frank W. Bailey, Russel Guest, *Above the Lines*, Grub Street, 1993.
Norman L.R. Franks, *Who Downed the Aces in WWI?*, Grub Street, 1996.
Norman L.R. Franks, Frank W. Bailey, *The Storks*, Grub Street, 1998.
Norman Franks, *Nieuport Aces of World War 1*, Osprey Publishing, 2000.
Norman L.R. Franks, Christophe Cony, *The French Air Service War Chronology 1914–1918*, Grub Street, 2001.
Vital Ferry, *French Aviation during the First World War, Histoire and Collections*, 2015
Réne Fonk, *Mes Combats*, (Ace of Aces), Doubleday & Co, 1967.
Jon Guttman, *SPAD VII Aces of World War 1*, Osprey Publishing, 2001.
Jon Guttman, *SPAD XII/XIII Aces of World War 1*, Osprey Publishing, 2002.
Jon Guttman, *SPA 124 Lafayette Escadrille*, Osprey Publishing, 2004.
Jon Guttman, *Groupe de Combat 12, 'Les Cigognes'*, Osprey Publishing, 2004.
Jon Guttman, *Balloon-Busting Aces of World War 1*, Osprey Publishing, 2005.
Jon Guttman, *Pusher Aces of World War 1*, Osprey Publishing, 2009.
Jon Guttman, *Resonnaisance and Bomber Aces of World War 1*, Osprey Publishing, 2015.
Jon Guttman, *Grim Reapers: French Escadrille 94 in World War I*, Aeronaut Books, 2016.
Jules Roy, Guynemer, L'Ange de La Mort, Albin Michel, 1986.
Jan Sumner, *The Kings of the Air: French Aces and Airmen of the Great War*, Pen and Sword Digital, 2015.
Jim Wilberg, Jean Navarre: *France's Sentinel of Verdun and the First French Fighter Pilots*, Aeronaut Books, 2010
Georges Thomas, *Guynemer, l'as des as*, Imp. de Vaugirard, 1917 (Gallica)

Periodicals

Aérojournal – No. 18, 26, 41, 47, 48, 49
Avions – Vols 1 to 235
Batailles Aeriennes – No. 74, 77, 81
Bulletin mensuel / Ligue aéronautique de France – 1917
Cross & Cockade (International) – Vols 1 to 44
Cross & Cockade (US) – Vols 1 to 26
HMP – No. 5, 6, 12/1992, 7/ 1995, 11/2003
La Vie aérienne : revue hebdomadaire paraissant le jeudi – 02/09/1919 (Gallica)
La Guerre aérienne illustrée – 1916, 1917, 1918, 1919 (Gallica)
Le Fana de l'Aviation – Vols 6 to 553
Le Monde illustré 1921 (Gallica)
Lotnictwo z szachownica – No. 1/ 2002
Modelář – No. 4/2004
Over the Front – Vols 1 to 31
Replic – No.9, 12, 43, 86, 194
Revi – No. 4, 50, 51
Squadron Signal No. 93 SPAD Fighters
Windsock International – Vols 1 to 31
Windsock Datafile – No. 8, 16, 20, 32, 36, 58, 68, 80, 94, 96, 108, 109, 135, 161, 167
Windsock Mini Datafile – 3, 5, 6
Windsock Datafile Special – Nieuport Fighters (Vols 1, 2), Breguet 14
Wing Masters – No. 61

Internet Resources

www.albindenis.free.fr
www.airwar.ru
www.as14-18.net
www.asoublies14-18.cosadgip.com
www.aviaww1.forum24.ru
www.bibert.fr
www.cieldegloire.com
www.deullin.fr
www.fandavion.free.fr
www.gallica.bnf.fr
www.memoiredeshommes.sga.defense.gouv.fr
www.mincbergr.net
www.navarre-jean.com
www.pages14-18.mesdiscussions.net
www.rene.fonck.free.fr
www.scalemodels.ru
www.theaerodrome.com
www.traditions-air.fr

Made in the USA
Coppell, TX
12 November 2026

11803126R00088